The Textile Matrix

Modeling, Measuring, and Mitigating Textile Microfiber Pollution

By Dr. Damon Rahimi Fard

ISBN 978-1-7646249-5-4

Disclaimer

This book is a scientific monograph on environmental science. The information presented is based on the author's research, analysis, and expertise. It is intended for academic, research, and educational purposes. While every effort has been made to ensure accuracy, the author and publisher make no warranties regarding the completeness or suitability of the content for any particular purpose. Readers should exercise professional judgment when applying any concepts, models, or recommendations contained herein.

A catalogue record for this book is available from the National Library of Australia.

ABOUT THE AUTHOR

Dr. Damon Rahimi Fard is an environmental materials specialist, civil engineer, and author based in Melbourne, Australia. He holds a Ph.D. in Civil Engineering, with a research specialization in water engineering, hydrology and municipal hydraulics. His academic and professional work focuses on the intersection of structural infrastructure and environmental forensics, examining the interaction between urban civil networks and microscopic polymer degradation pathways.

Dr. Rahimi Fard's research addresses the physical, thermodynamic, and mechanical mechanisms of synthetic material failure within hydrological systems. He is the author of the monograph *The Synthetic Biosphere* (2026), which introduced frameworks for environmental gatekeeping and engineered molecular erasure. In *The Textile Matrix*, he builds on this work, utilizing materials science, data engineering, and civil infrastructure controls to model, measure, and mitigate polymer and nanoplastic leakage at the source. This focused study was necessary because *The Synthetic Biosphere* addressed plastic pollution across a broad range of environmental domains, leaving room for a dedicated deep dive into textile microfiber pollution specifica

Contents

Chapter 1: The Microplastic Paradigm and the Textile Matrix: Proliferation, Release Pathways, and the Imperative for Source Reduction

Introduction

In the Anthropocene, the proliferation of microplastics represents one of the most insidious and far-reaching anthropogenic alterations to Earth's biogeochemical cycles. This monograph provides a dedicated, technically rigorous analysis of one of the most significant yet addressable vectors of microplastic contamination originating from the textile sector.

This chapter provides the essential scientific foundation for the analyses that follow by:

- Contextualizing the exponential growth of synthetic polymer production and its integration into every domain of modern civilization.
- Establishing a rigorous taxonomy of plastic debris by dimensional scale, morphology, and lifecycle origin.
- Examining the textile matrix in depth: global production dynamics, fiber categories, and fragmentation propensity.
- Quantifying and characterizing the two dominant release pathways—pre-consumer manufacturing emissions and domestic laundering.

1.1 The Ascendancy of Synthetic Polymers: Production Dynamics and Industrial Ubiquity

Global plastic production has sustained an unrelenting upward trajectory. By the mid-2020s, annual output had reached approximately 450–460 million metric tons (MMT). Projections from the OECD and PlasticsEurope indicate that this figure could surpass 600 MMT by 2040 under business-as-usual conditions, carrying profound implications for planetary boundaries and Earth system integrity.

Synthetic polymers have become indispensable to modern existence. Their exceptional versatility, mechanical durability, chemical inertness, low production cost, and economic accessibility have driven integration across virtually every domain of contemporary infrastructure and daily life—from commercial transportation networks and civil engineering to advanced electronics, performance apparel, medical devices, and household products.

Industrial Exploitation and Sector-Specific Utilities

Healthcare and Pharmaceuticals: Plastics are indispensable in modern medicine, supplying the sterility, flexibility, and biocompatibility required for essential devices such as single-use syringes, intravenous systems and tubing, blood bags, surgical gloves, catheters, prosthetic implants, and advanced drug delivery systems. Pharmaceutical packaging relies almost universally on lightweight, chemically inert polymer containers engineered to shield sensitive compounds from moisture, light, oxygen, and microbial contamination. The COVID-19 pandemic dramatically accelerated global demand for medical-grade plastics, underscoring their critical role in healthcare delivery.

Food Preservation and Supply-Chain Logistics: Polymer-based protective films, hermetic containers, modified atmosphere packaging, and beverage bottles form the backbone of the global food distribution network. These solutions substantially reduce post-harvest losses—often by 30–40% for perishable goods—while lowering transportation mass compared with traditional glass or metal alternatives.

Civil Infrastructure and Transportation Systems: Engineering-grade polymers are now essential in automotive components, fuel systems, plumbing networks, thermal insulation, roofing membranes, and electrical conduit. The construction sector alone accounts for roughly one-fifth of global plastic consumption, with long-term implications for waste management as buildings reach end-of-life.

Technical Textiles and Performance Apparel: Synthetic fibers deliver superior mechanical performance, durability, and functional properties (moisture wicking, thermal regulation, UV protection) that natural fibers struggle to match at comparable cost. This dominance has simultaneously amplified microfiber shedding, establishing textiles as a major vector for environmental microplastic contamination.

1.2 A Taxonomy of Plastic Debris: Dimensional Scales, Morphological Features, and Lifecycle Origins

Environmental plastic debris is systematically classified according to both dimensional criteria and lifecycle provenance. This taxonomy provides the essential framework for source attribution, transport modeling, risk assessment, and the design of targeted remediation interventions.

1.2.1 Primary versus Secondary Microplastics

Primary microplastics are particles deliberately manufactured at microscopic dimensions for specific commercial or industrial functions—exfoliating microbeads (largely phased out through regulatory action), industrial abrasives, and pre-production polymer pellets (nurdles). These particles enter the environment directly through product use, manufacturing discharge, or accidental spillage.

Secondary microplastics, by contrast, arise from the progressive fragmentation of larger plastic items through solar ultraviolet radiation, oxidative weathering, mechanical abrasion, thermal cycling, and biological activity. This category constitutes the overwhelming majority—estimated at 70–90%—of environmental microplastic burdens and originates from mismanaged municipal litter, single-use packaging, agricultural films, tire wear particles, and, critically for this monograph, degraded synthetic textiles.

1.2.2 Dimensional and Morphological Classification Framework

Regulatory and scientific consensus, synthesized from the European Union plastics strategy, AATCC, The Microfibre Consortium, and ISO, has established unified size-based nomenclature. Microplastics are formally defined as water-insoluble polymeric particles whose longest dimension does not exceed 5 mm. Particle size governs environmental transport, bioavailability, cellular uptake, and toxicological mechanisms. Nanoplastics (<1 μm) can cross biological membranes, including the blood-brain and placental barriers. Microplastics (1 μm–5 mm) are readily ingested across trophic levels.

Table 1.1: Environmental Classification of Plastic Debris by Dimensional Scale

Debris Category	Dimensional Range	Typical Morphology & Sources	Dominant Transport & Fate
Nanoplastics	< 1 μm	Colloidal fragments; high intracellular penetration	Cross biological membranes; ultra-high surface reactivity
Microplastics	1 μm - 5 mm	High-aspect-ratio microfibers, nurdles, fragmented packaging, textile debris	Ubiquitous via wind/currents; high bioavailability
Mesoplastics	5 mm - 25 mm	Industrial nurdles, bottle caps, weathered shards, fishing gear	Intermediate transport; often stranded on shorelines
Macroplastics	25 mm - 1 m	Consumer bags, containers, agricultural films, derelict fishing gear	Surface/nearshore accumulation; entanglement hazards
Megaplastics	> 1 m	Ghost nets, agricultural sheeting, large waste items, derelict vessels	Long-range oceanic transport via gyres; habitat disruption

1.2.3 Chemical Profiles and Global Polymer Distribution

The polymer composition of recovered textile microplastics closely mirrors global synthetic fiber production volumes. Polyethylene terephthalate (PET) typically accounts for 60–70% of identified microfibers, followed by polyamides and polyacrylonitrile. Each polymer class displays distinct environmental behaviors governed by its molecular structure and susceptibility to specific degradation mechanisms.

- Acrylic fibers (polyacrylonitrile): Particularly susceptible to photo-oxidative chain scission under solar UV, leading to rapid surface micro-cracking.
- Polyamides (Nylon-6 and Nylon-6,6): Heightened vulnerability to hydrolytic degradation in warm or acidic marine conditions due to high moisture regain.
- Polyester (PET): Slower but persistent hydrolytic and photo-oxidative attack; maintains structural integrity for decades in marine environments.
- Polypropylene (PP) and Polyethylene (PE): Highly susceptible to photo-oxidation but relatively resistant to hydrolysis; low density causes initial flotation followed by biofouling-induced sinking.

1.3 The Textile Matrix: Global Production Dynamics, Fiber Categorization, and Fragmentation Propensity

While microplastics originate from many sources, textile-derived microfibers constitute a singularly significant and growing contributor. Independent studies attribute approximately 35% of primary microplastics entering the global ocean to apparel and home textile laundering (Boucher & Friot, 2017). The textile sector is therefore not merely one source among many—it is a structurally dominant vector whose mitigation offers disproportionate leverage for reducing the overall microplastic burden.

1.3.1 Global Production Context and Fiber Categories

Global fiber production has more than doubled since the turn of the century. Output reached approximately 124 million tonnes in 2024 and is projected to attain 150–160 million tonnes by 2030 under current trajectories. Synthetic fibers, predominantly petroleum-derived, now constitute 65–70% of total volume. China alone accounts for roughly 65% of global synthetic fiber output.

The global textile complex sheds fibrous fragments from three principal feedstock categories:

1. Synthetic Fibers (PET, PA, PAN): Non-biodegradable and reliant on petrochemical feedstocks. Dominant in high-shedding applications such as fleece, activewear, and technical textiles.
2. Man-Made Cellulosic Fibers (viscose, modal, lyocell): Derived from renewable biomass but subjected to intensive chemical processing that can retard biodegradation and introduce legacy pollutants.
3. Natural Fibers (cotton, wool, hemp, etc.): Inherently biodegradable under favorable aerobic conditions but often carry agricultural residues and synthetic surface finishes that complicate their environmental profile.

1.3.2 Nanoplastics as the Terminal Fragmentation Product of Textile Fibers

Section 1.2.2 introduced nanoplastics as particles smaller than 1 μm capable of crossing biological membranes. This section examines their formation specifically from textile fibers — a mechanism, morphology, and surface chemistry that differ substantially from pellet-origin or rigid-fragment nanoplastics — and traces the implications through to the toxicological mechanisms detailed in Chapter 4.

Textile microfibers do not remain at the micrometer scale indefinitely. Through the combined action of ultraviolet photo-oxidation, mechanical abrasion during laundering and environmental transport, and biological activity within the plastisphere, these fibers undergo progressive fragmentation that ultimately yields nanoplastics (<1 μm). This terminal fragmentation stage represents the most consequential and least examined link in the textile pollution lifecycle, completing the chain from domestic emission to cellular-level infiltration. The fragmentation kinetics are modelled in detail in Chapter 3 as part of the domestic laundering emission engine.

The fragmentation kinetics of textile fibers are distinct from those of primary industrial pellets. Synthetic textile fibers (particularly polyester and polyamide) possess high aspect ratios and relatively low surface crystallinity, rendering them susceptible to chain scission. Ultraviolet radiation generates carbonyl and hydroxyl groups along the polymer backbone, reducing molecular weight and creating surface micro-cracks that serve as initiation sites for mechanical failure. During domestic laundering, hydrodynamic shear, elevated temperature (30–60°C), and surfactant action accelerate surface erosion. Literature reports indicate that a typical 300–500 g polyester garment sheds 0.5–5 mg of microfibers per wash cycle (Napper & Thompson, 2016), with an increasing proportion of the released mass entering the sub-micron range after repeated laundering and environmental weathering (Hernandez et al., 2019). In aquatic systems, continuous exposure to wave action, sediment abrasion, and biofilm colonization sustains the fragmentation cascade. The process follows a power-law distribution in which particle number increases accelerate as average particle size decreases, ultimately generating a polydisperse population of nanoplastics that retain elongated or irregular morphologies even below 1 μm.

A critical and underappreciated distinction lies in surface chemistry and eco-corona formation. An eco-corona is the layer of biomolecules (primarily proteins, polysaccharides, and lipids) that rapidly adsorbs onto the surface of a particle upon contact with an aqueous environment, fundamentally altering its biological identity and interaction with organisms. Primary nanoplastics derived from pre-production pellets typically present relatively homogeneous surfaces once initial weathering has occurred. In contrast, nanoplastics originating from textile fibers carry a legacy of manufacturing additives, dyes, optical brighteners, antistatic agents, and residual catalysts originally incorporated into the yarn or fabric. These embedded compounds create heterogeneous surface domains that alter protein adsorption kinetics and eco-corona composition, differentiating them entirely from simpler particles originating from packaging, agricultural films, or tire wear. The protein corona formed around fiber-derived nanoplastics is therefore chemically and structurally distinct from that surrounding pellet-derived particles at equivalent mass concentrations. This variance directly influences opsonization, cellular recognition, and uptake pathways, allowing fiber-origin nanoplastics to present surface-bound additives directly to cellular membranes and generate localized, high-concentration chemical exposure zones.

The transition below 1 μm fundamentally alters the toxicological relevance of textile-derived contamination. At the fiber scale (typically 10–50 μm diameter), particles remain largely confined to the gastrointestinal tract or respiratory surfaces. Once fragmentation reaches the sub-micron regime, these particles acquire the capacity to cross biological barriers previously considered impermeable. Nanoplastics penetrate the alveolar-capillary interface, enter systemic circulation, and subsequently breach the blood-brain and placental barriers (Amato-Lourenço et al., 2021; Ragusa et al., 2021). This capability directly connects the physical lifecycle of consumer-phase laundering and atmospheric emissions (Sections 1.4 and 1.5) to the somatic infiltration, transgenerational exposure, oxidative stress, inflammasome activation, and lysosomal rupture pathways detailed in Chapter 4. Current literature frequently treats nanoplastics generically without origin-specific differentiation. This obscures the massive textile contribution to the nanoscale burden, underestimating both exposure magnitude and mechanistic specificity in risk assessment and regulatory prioritization.

Because textile fibers constitute one of the largest mass fluxes of secondary microplastics into the environment and undergo continuous fragmentation throughout their residence time, textile-derived nanoplastics likely represent a dominant and chemically distinct fraction of environmentally relevant nanoplastics. Their retained elongated morphologies, distinct surface chemistry, and embedded additive profiles differentiate them from particles originating from packaging, agricultural films, or tire wear. This distinction carries consequences for exposure modeling, toxicological risk assessment, and regulatory prioritization.

Current microfiber mitigation technologies — washing machine filters, advanced wastewater treatment, and yarn engineering — primarily target the micrometer scale. While effective at reducing larger microfibers, they exert limited control over the continuous generation of nanoplastics through ongoing environmental fragmentation. Addressing the terminal nanoscale stage requires complementary interventions: accelerated degradation technologies operating below 1 μm, surface modification strategies that suppress fragmentation propensity during fiber production, and regulatory frameworks that explicitly incorporate nanoscale emissions into textile life-cycle assessments.

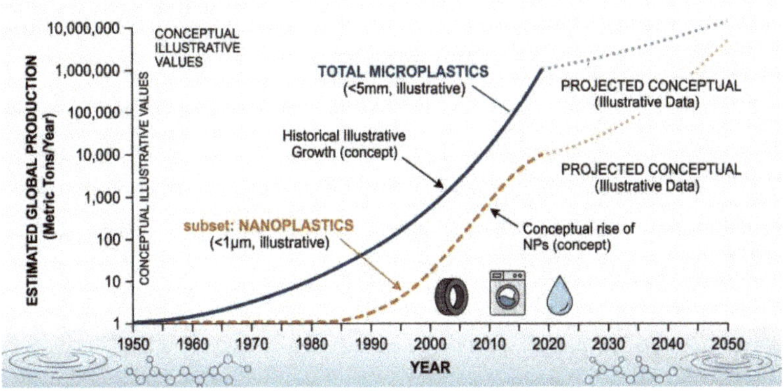

Figure 1.1: Historical and Projected Global Microplastic and Nanoplastic Production

1.4 Pre-Consumer Emission Hotspots within the Textile Manufacturing Chain

Although consumer laundering is widely recognized as a dominant release pathway, substantial quantities of microplastics are generated during pre-consumer manufacturing stages. These emissions represent a critical intervention point for source reduction, as they occur before garments enter consumer circulation and can be captured through industrial engineering controls and filtration systems.

1.4.1 Fiber Extrusion and Yarn Spinning Operations

During melt-spinning, high-purity polymer melts are extruded under high pressure through micro-scale spinneret orifices. Subsequent drawing, texturizing, carding, and combing operations subject the filaments to continuous high-velocity friction against metal guides and ceramic components. This abrasive contact generates elevated concentrations of pre-consumer microplastic dust. These manufacturing-derived fragments are often embedded within the yarn structure and are not released until the garment undergoes its first domestic wash cycles, creating a concentrated "first-flush" emission pulse into municipal wastewater systems.

1.4.2 Weaving, Knitting, and Industrial Cutting Processes

Weaving and knitting subject yarns to high-speed cyclic bending, interlacing, and abrasion. During garment assembly, laser or mechanical blade cutting shears across thousands of yarn loops, releasing substantial volumes of loose fiber fragments and cutting debris. When pre-consumer garments bypass industrial pre-washing and filtration, these fragments are rapidly mobilized during the consumer's initial domestic wash cycles.

1.5 Domestic Laundering: A Critical Consumer Release Vector

Routine domestic laundering constitutes the dominant consumer-phase release mechanism for textile microfibers. The International

Union for Conservation of Nature (IUCN) has attributed approximately 35% of primary microplastics entering the global ocean to apparel and home textile laundering (Boucher & Friot, 2017). The volume and morphology of released fibers are governed by the complex interplay of fiber/yarn properties, fabric structural design, and washing-machine operational parameters.

1.5.1 Fiber and Yarn Properties

- Staple versus Continuous Filament: Yarns spun from short staple fibers shed far more profusely than continuous-filament yarns because unconstrained fiber ends readily slip from the yarn body under agitation.
- Yarn Twist Multiplier: High twist increases internal friction and cohesion between fibers, suppressing shedding; low-twist yarns are characteristically hairy and prone to fiber release.
- Combing versus Carding: Combed yarns undergo removal of short fibers and parallel alignment, yielding low hairiness and superior shedding resistance compared with carded yarns.

1.5.2 Fabric Structural Design

Knitted structures (jersey, interlock, fleece) feature interlocking loops that stretch and abrade under washing forces, promoting high fiber release. Woven fabrics, with perpendicular warp/weft yarns under high tension, physically constrain fibers and markedly reduce shedding. High-density, low-porosity constructions trap loose fibers internally. Mechanical raising, napping, or shearing (common in polar fleeces) deliberately breaks filaments to create pile, generating large populations of loose fragments that shed rapidly during initial wash cycles.

1.5.3 Washing-Machine Operational Parameters

- Top-loading machines with central vertical agitators impose severe mechanical stress, releasing up to seven times more microfibers than front-loading tumble machines.

- Elevated temperatures (>40°C) swell fibers and accelerate hydrolytic degradation; cold-water washing (<30°C) is a simple, effective mitigation strategy.
- Surfactants and high-pH formulations can increase fiber release; liquid detergents and optimized formulations generally perform better.
- Washing only full loads and selecting gentle cycles reduces mechanical friction per garment.

1.6 Emerging Policy and Research Developments

Since the publication of foundational assessments such as Boucher & Friot (2017), the scientific and policy landscape has evolved significantly. The ongoing negotiations toward a legally binding global plastics treaty under the United Nations Environment Assembly have elevated textile microfibers as a priority category for source reduction. Parallel regulatory developments—including the European Union's Ecodesign for Sustainable Products Regulation (ESPR) and mandatory Digital Product Passports—now explicitly target microfiber shedding through design requirements and extended producer responsibility frameworks. These policy instruments create new leverage points for upstream intervention that complement the technical mitigation strategies detailed in subsequent chapters.

References

1. Geyer, R., Jambeck, J. R., & Law, K. L. (2017). Production, use, and fate of all plastics ever made. Science Advances, 3(7), e1700782.

2. PlasticsEurope. (2025). Plastics – the Fast Facts 2025. PlasticsEurope Market Research Group.

3. Textile Exchange. (2025). Materials Market Report 2025. Textile Exchange.

4. OECD. (2022). Global Plastics Outlook: Policy Scenarios to 2060. OECD Publishing, Paris. https://doi.org/10.1787/aa1edf33-en

5. European Commission. (2018). A European Strategy for Plastics in a Circular Economy. COM(2018) 28 final.

6. The Microfibre Consortium. (n.d.). Glossary of terms relating to microfibres and fibre fragments. https://www.microfibreconsortium.com/glossary

7. Boucher, J., & Friot, D. (2017). Primary microplastics in the oceans: a global evaluation of sources. IUCN.

8. Henry, B., Laitala, K., & Klepp, I. G. (2019). Microfibres from apparel and home textiles. Sustainable Production and Consumption, 20, 263–275.

9. Napper, I. E., & Thompson, R. C. (2016). Release of synthetic microplastic fibres from domestic washing machines. Marine Pollution Bulletin, 112(1-2), 39–45.

10. European Commission. (2022/2024). Ecodesign for Sustainable Products Regulation (ESPR) and Digital Product Passport framework. Brussels. (Updated regulatory developments targeting microfiber shedding).

11. United Nations Environment Assembly. (2022–ongoing). Intergovernmental Negotiating Committee on Plastic Pollution (Global Plastics Treaty negotiations). Nairobi. (Elevating textile microfibers as a priority category for source reduction).

12. Hernandez, E., et al. (2019). Polyester textiles as a source of microplastics from households: A mechanistic study to understand microfiber release during washing. Environmental Science & Technology, 53(21), 12579–12588.

13. Amato-Lourenço, L. F., et al. (2021). Presence of airborne microplastics in human lung tissue. Journal of Hazardous Materials, 416, 126124.

14. Ragusa, A., et al. (2021). Plasticenta: First evidence of microplastics in human placenta. Environment International, 146, 106274.

15. Galloway, T. S., et al. (2017). Interactions of microplastic debris throughout the marine ecosystem. Nature Ecology & Evolution, 1, 0116.

Chapter 2: Environmental Transport, Degradation, Toxicological Risks, and Strategic Imperatives for Engineered Biospheric Remediation

Introduction

Chapter 1 established the principal sources and release pathways of textile-derived microplastics, demonstrating that the textile sector constitutes a structurally dominant vector whose mitigation offers disproportionate leverage for reducing the global microplastic burden. Having delineated where these persistent contaminants originate and how they enter environmental compartments, this chapter traces their subsequent journey through Earth's surface systems.

The analysis proceeds through four interconnected domains:

4. Environmental Transport and Fate — the physical pathways (aquatic, atmospheric, sedimentary, indoor) that govern dissemination, vertical flux, and accumulation hotspots.
5. Thermodynamic and Mechanistic Pathways of Polymer Aging and Degradation — the photo-oxidative, hydrolytic, and mechanical processes that progressively fragment macro- and microplastics into nanoplastics.
6. Toxicological Vectors and Systemic Risks — the physical, chemical, and biological mechanisms through which microplastics and their associated co-pollutants threaten ecological integrity, biodiversity, and human health.
7. Strategic Imperatives for Source Reduction, Policy, and Molecular Erasure — the integrated upstream, midstream, and downstream interventions required to transition toward a circular, non-toxic textile economy and ultimately achieve

engineered molecular erasure of synthetic contaminants from the biosphere.

This chapter bridges the source-focused analysis of Chapter 1 with the remediation technology selection, machine-vision monitoring, and circular business model innovations developed in this monograph.

2.1 High-Performance Technical Textiles Deployed in Open Environmental Compartments

Technical textiles constitute a specialized class of high-performance materials engineered primarily for functional attributes rather than aesthetics. Classified by Messe Frankfurt into twelve sectors (Agrotech, Buildtech, Clothtech, Geotech, Hometech, Indutech, Medtech, Mobiltech, Packtech, Protech, Sporttech, and Oekotech), these materials frequently employ high-tenacity synthetic polymers specifically selected to withstand extreme physical and environmental stresses (Messe Frankfurt). Because many technical textiles are deployed directly within open agricultural, civil-engineering, marine, and industrial environments, their continuous weathering constitutes a significant yet under-regulated source of environmental microplastics.

2.1.1 Agrotextiles (Agrotech)

Agrotextiles encompass weed-control ground covers, greenhouse shade nets, crop protection films, harvesting nets, and agricultural twines. Typically manufactured from polypropylene (PP) or high-density polyethylene (HDPE), these materials are spread across expansive fields and exposed to intense solar UV radiation, elevated temperatures, wind shear, and agrochemicals. Over successive growing seasons, photo-oxidative embrittlement and mechanical stress cause fragmentation directly into topsoils, compromising soil structure, reducing water retention, and introducing microplastics into terrestrial food webs. The transition to biodegradable alternatives (PLA, PHA, PBS) is a priority recommendation for reducing long-term soil contamination.

2.1.2 Geotextiles (Geotech)

Geotextiles are employed in civil engineering for soil stabilization, road reinforcement, filtration, drainage, erosion control, and coastal protection. Non-woven needle-punched or woven high-tenacity sheets are buried in direct contact with soil, groundwater, or river systems. Decades of service subject the fibers to compaction, shear friction from soil particles, hydraulic forces, freeze-thaw cycling, and microbial activity. The resulting microplastics are transported into sub-surface aquifers, rivers, and estuaries, representing a persistent hidden pathway of secondary contamination that is often overlooked in surface-water monitoring programs.

2.1.3 Marine and Fishing Technical Textiles (Indutech / Protech)

Commercial fishing nets, aquaculture cages, mooring ropes, sails, and protective covers rely predominantly on high-tensile polyamide (nylon) and polypropylene. Continuous exposure to wave action, mechanical drag, salinity, UV radiation, and winch-drum friction shears microfibers directly into the water column. Abandoned, lost, or otherwise discarded fishing gear (ALDFG, or "ghost gear") persists indefinitely, slowly fragmenting into millions of highly mobile secondary microplastics that present direct ingestion hazards to marine life across pelagic and benthic zones. Biodegradable net mandates and ALDFG recovery programs are priority policy recommendations supported by growing international consensus.

2.2 Environmental Transport and Fate: Aquatic Dynamics, Sedimentary Sinks, and Atmospheric Pathways

Once released into the environment, microplastics are governed by a complex array of transport forces across multiple physical media (atmosphere, freshwater, marine, terrestrial, and cryospheric). Understanding these pathways is essential for modeling exposure risks, identifying accumulation hotspots, and designing effective remediation systems.

2.2.1 Synthetic Textile Emissions and Aquatic Fate

Synthetic textiles are non-biodegradable under natural conditions and embody substantial embodied energy and greenhouse-gas emissions. End-of-life circularity remains severely constrained: only 15–20% of post-consumer textile waste enters recycling streams in most jurisdictions. Once released into water systems, polymer density governs initial buoyancy and subsequent vertical transport.

Table 2.1: Comparative Density and Aquatic Behavior of Common Synthetic Polymers

Polymer	Acronym	Density (g/cm^3)	Buoyancy in Seawater	Primary Environmental Fate
Polypropylene	PP	0.89–0.91	Positively buoyant	Surface-layer transport; eventual biofouling and sinking
Polyethylene	PE	0.91–0.97	Positively buoyant	Surface films; rapid biofouling-induced sinking
Polyvinyl Chloride	PVC	1.30–1.58	Negatively buoyant	Rapid benthic sedimentation; deep-ocean accumulation
Polyamide 6/6,6	PA6/PA6,6	1.13–1.15	Negatively buoyant	Intermediate to benthic water-column distribution
Polyethylene Terephthalate	PET	1.38–1.40	Negatively buoyant	Benthic and coastal sediment sinks

Low-density polymers (PP, PE) initially float but quickly acquire microbial biofilms that increase their effective density, driving eventual sinking. This biofouling-mediated vertical flux ensures microplastics permeate the entire water column and sediment record—from surface waters to hadal trenches. The "plastisphere" concept (Zettler et al., 2013) is central to bioremediation strategies, as plastic particles serve as novel substrates for microbial colonization and potential pathogen transport.

2.2.2 Atmospheric Microplastics: Inhalation Pathways and Urban Exposure Dynamics

Microplastics are now recognized as a significant constituent of atmospheric particulate matter (PM), presenting direct inhalation and secondary ingestion risks. Urban infrastructure can inadvertently amplify airborne dissemination through aerated wastewater tanks, decorative fountains, and building ventilation exhausts. Atmospheric microplastics (typically 1–100 μm) possess highly reactive surfaces that readily adsorb heavy metals, pharmaceutical residues, and persistent organic pollutants. Subsequent environmental aging further enhances their capacity to concentrate and later desorb hazardous co-pollutants.

2.2.3 Indoor Air Quality and Human Exposure

Contemporary urban populations spend up to 90% of their time indoors. Indoor dust matrices serve as reservoirs for substantial loads of microplastic fibers shed from synthetic carpets, upholstery, clothing, bedding, and forced-air ventilation systems. A landmark Paris study recorded indoor microplastic concentrations of 1–60 particles/m^3, compared with 0.3–1.5 particles/m^3 in contemporaneous outdoor air (Dris et al., 2017). This order-of-magnitude elevation underscores the elevated inhalation hazard within modern, well-insulated buildings and highlights the need for improved indoor air filtration and source control in residential and commercial settings.

2.3 Thermodynamic and Mechanistic Pathways of Polymer Aging and Degradation

The conversion of high-strength synthetic textiles into microscopic fragments is governed by interconnected thermodynamic, chemical, and physical degradation mechanisms. Polymer aging modifies molecular weight, crystallinity, surface chemistry, and mechanical properties, rendering fibers progressively susceptible to fragmentation.

2.3.1 Photo-Oxidative Degradation

Photo-oxidation constitutes the dominant chemical degradation pathway for polymers exposed to solar radiation. Ultraviolet photons are absorbed by the polymer backbone, generating carbon-centered radicals through homolytic bond cleavage. These radicals rapidly combine with atmospheric oxygen to form peroxy radicals, which propagate a chain reaction producing hydroperoxides and additional molecular radicals. The resulting reduction in molecular weight and increase in crystallinity produce a brittle material characterized by surface micro-cracks, deep fissures, and pitting—readily observable via FTIR spectroscopy through a rising carbonyl index (CI). These defects function as stress concentrators, dramatically lowering the energy threshold required for mechanical fragmentation (Andrady, 2017).

2.3.2 Hydrolytic Degradation

Hydrolysis involves nucleophilic cleavage of ester, amide, or ether linkages by water molecules. The process is strongly modulated by temperature, pH, salinity, and polymer crystallinity. Polyester (PET) ester linkages undergo slow hydrolytic scission in marine environments. Polyamides (Nylon) amide linkages are highly susceptible to acid-catalyzed hydrolysis due to high moisture regain (4–5% by weight). Hydrolytic degradation is a key mechanism exploited in enzymatic and chemical recycling technologies and contributes to the long-term breakdown of synthetic fibers in aquatic systems.

2.3.3 Mechanical Friction and Cyclic Fatigue

Aged fibers containing pre-existing micro-cracks experience accelerated fatigue failure under cyclic mechanical loading—whether from domestic laundering agitation, continuous marine wave action, or repeated flexing in technical textile applications. Micro-cracks propagate rapidly across the filament cross-section under bending or tensile stress until catastrophic fracture occurs. A single physical impact event can generate hundreds of microscopic fragments from a single embrittled filament, amplifying secondary microplastic loads in high-energy environments.

2.4 Toxicological Vectors and Systemic Risks: Ecological and Human Health Dimensions

Microplastics constitute a multifaceted environmental hazard through physical, chemical, and biological mechanisms that threaten ecological integrity, biodiversity, and human health.

2.4.1 Aquatic Ecosystems and Trophic Transfer

Aquatic environments bear the most immediate burden. Millions of tonnes of microplastics enter marine and freshwater systems annually. Because particle dimensions overlap with those of plankton and other basal food web components, organisms across trophic levels ingest them readily. Subsequent biomagnification across trophic levels imperils marine biodiversity, disrupts commercial fisheries, and undermines the economic resilience of coastal communities. The plastisphere concept—plastic particles serving as substrates for pathogenic and invasive microbial communities (Zettler et al., 2013)—adds a biological hazard dimension with implications for disease dynamics and ecosystem stability.

2.4.2 Chemical Hazard Vectors

Microplastics function as mobile concentrators of environmental contaminants. Their large surface-area-to-volume ratio enables adsorption and concentration of hydrophobic pollutants from surrounding media. Simultaneously, they leach intrinsic manufacturing additives—phthalate plasticizers, bisphenol A (BPA), flame retardants, and other known endocrine-disrupting chemicals (EDCs)—that interfere with hormonal signaling, reproductive success, and cellular homeostasis in exposed organisms (Cole et al., 2011).

2.4.3 Biological Hazard Vectors

The same high surface area facilitates colonization by pathogenic microorganisms, toxic algal strains, and invasive species. Plastic particles thereby serve as vectors for disease transmission and ecological disruption. Empirical evidence increasingly documents microplastic presence in human blood, placental tissue, lung parenchyma, and other tissues, raising urgent questions regarding long-term cellular toxicity, inflammatory cascades, and potential neurodevelopmental or carcinogenic outcomes.

2.5 Synthesis: Material Properties, Exposure Profiles, and the Research Landscape

To integrate the physical characteristics, chemical attributes, degradation pathways, and environmental release behaviors across fiber types and technical textile sectors, the following tables synthesize key performance and emission metrics.

Table 2.2: Physical Properties and Environmental Persistence of Common Textile Fibers

Fiber / Polymer	Common Applications	Density (g/cm^3)	Tensile (g/den)	Moisture Regain (%)	Primary Degradation Pathway
Cotton	Apparel, bedding, denim	1.54	3.0–5.0	7–8	Enzymatic hydrolysis (cellulases)
Wool	Heavy coats, winter knitwear	1.31	1.0–1.7	13–16	Microbial proteolysis & enzymatic decay
Polyester (PET)	Activewear, fleeces, geotextiles	1.38	4.0–7.0	0.4	Photo-oxidation + slow hydrolysis
Polyamide (Nylon)	Hosiery, swimwear, fishing nets	1.14	4.5–7.0	4–5	Acid-catalyzed hydrolysis + photo-oxidation
Acrylic (PAN)	Winter sweaters, outdoor carpets	1.17	2.0–3.5	1–2	Photo-oxidative degradation + thermal wear
Polypropylene (PP)	Agrotextiles, hygiene products	0.91	3.5–6.0	<0.1	Rapid photo-oxidation (requires stabilizers)

Table 2.3: Microplastic Emission Profiles and Operational Parameters of Technical Textile Sectors

Technical Sector	Primary Polymer(s)	Deployment Environment	Primary Weathering Stressors	Microplastic Release Mechanism	Mitigation Strategies
Agrotextiles (Agrotech)	PP, HDPE	Exposed agricultural topsoils	Solar UV-A/B, heat, agrochemicals	Brittle photo-oxidative fragmentation	Organic UV absorbers; transition to PLA/PHA
Geotextiles (Geotech)	PET, PP	Buried sub-surface soils	Compaction, soil friction, groundwater	Mechanical abrasion + hydraulic detachment	High-tenacity continuous-filament weaves
Marine Textiles (Indutech)	PA (Nylon), PP	Pelagic & coastal marine zones	Wave action, salinity, drag, UV	Mechanical tearing, cyclic fatigue, ghost-gear	Biodegradable net mandates; ALDFG recovery
Mobiltech (Automotive)	PET, PA	Vehicle cabins & engine compartments	Friction, vibration, thermal cycling	Surface wear under passenger motion	Protective coatings; high-twist combed yarns

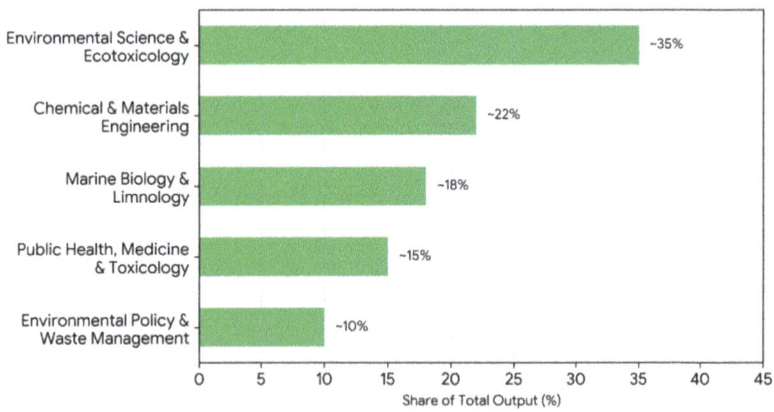

Figure 2.1: Academic Publication Distribution by Subject Domain

Note: Academic Publication Distribution by Subject Domain. Source: Scopus, keyword "microplastic," articles and reviews, January 2024.

Percentages are illustrative of relative research emphasis and approximate; exact values vary with search parameters.

2.6 Strategic Imperatives for Source Reduction, Policy, and Molecular Erasure

Mitigation demands an integrated, lifecycle-oriented strategy that simultaneously addresses upstream material innovation, midstream engineering controls, and downstream policy frameworks.

Table 2.5: Systemic Remediation Framework

Intervention Pillar	Core Strategies	Expected Outcomes
Materials Redesign	Next-generation bio-based polymers; non-shedding textile architectures; SSbD frameworks	Reduced shedding; accelerated end-of-life mineralization; lower fossil-carbon dependence
Engineering Upgrades	Mandatory micro-filtration on washing machines; advanced tertiary treatment at WWTPs; industrial pre-consumer capture	Capture of >90% of laundry-derived microfibers; reduced effluent loads
Policy & Governance	Extended Producer Responsibility (EPR); design-for-environment mandates; international treaties; green public procurement	Internalization of end-of-life costs; accelerated innovation; source reduction at industrial scale

2.6.1 Upstream, Midstream, and Downstream Policy Implementations

- Upstream Material Innovation: Accelerate the commercialization of certified bio-based and biodegradable polymers (PLA, PHA, PBS, PHB) for high-emission sectors such as agrotextiles and technical textiles. Invest in research on inherently low-shedding fiber architectures and surface modifications.
- Yarn and Fabric Engineering: Optimize manufacturing parameters—using continuous-filament yarns, elevated twist multipliers, combed over carded structures, and transitioning from loose knitted fleeces to high-density woven geometries—to suppress shedding at the source.

- Advanced Eco-Friendly Surface Coatings: Develop and deploy durable, non-toxic biopolymer finishes (pectin-, chitosan-, or cellulose-based coatings) that mechanically lock surface fibers into the yarn body, achieving up to 90% reduction in mechanical shedding while preserving breathability.
- Pre-Consumer Source Capture: Mandate high-efficiency particulate (HEPA) filtration and multi-stage industrial wastewater microfiltration systems in all spinning, weaving, knitting, and garment-assembly facilities.
- Consumer Laundering Controls: Standardize integrated high-efficiency microfiber filters on all new residential and commercial washing machines, capable of intercepting up to 90% of effluent microfibers.
- Extended Producer Responsibility (EPR) Frameworks: Implement comprehensive EPR legislation requiring fashion brands and technical-textile manufacturers to finance collection, sorting, and chemical recycling infrastructure.
- Consumer Education and Behavioral Interventions: Develop and disseminate evidence-based guidance on low-shedding laundry practices (cold water, full loads, liquid detergent, line drying) and promote circular consumption models (rental, resale, repair).

2.7 Emerging Frontiers: Climate Interactions, Digital Traceability, and the Path to Net-Zero Microplastic Textiles

2.7.1 Climate-Microplastic Feedback Loops

The interaction between microplastic pollution and global climate systems is an area of rapidly growing concern. Microplastics in the ocean surface microlayer can alter the physical properties of the sea surface, potentially influencing air-sea gas exchange and the formation of sea spray aerosols that act as cloud condensation nuclei. In terrestrial systems, microplastic contamination of agricultural soils can modify soil structure, water retention, and microbial community composition, with potential consequences for soil carbon sequestration and greenhouse gas emissions (N_2O and CH_4). Furthermore, the photo-oxidative degradation of microplastics under intensified UV radiation in a warming climate may accelerate fragmentation rates, creating a positive feedback loop that warrants urgent research attention.

2.7.2 Digital Traceability and Blockchain-Enabled Material Passports

The emergence of Digital Product Passports (DPPs) and blockchain-based material traceability systems offers a transformative opportunity to close the information gap that currently hampers effective source attribution and circular design. By embedding immutable records of polymer type, additive chemistry, manufacturing provenance, and use-phase history directly into garments, these systems enable automated sorting at end-of-life, precise allocation of EPR fees based on actual environmental performance, and real-time verification of recycled content claims.

2.7.3 Thermodynamic and Economic Limits of Zero-Shedding Design

While the technical pathways to dramatically reduced shedding are becoming clearer (high-twist continuous filaments, advanced surface coatings, and bio-based alternatives), fundamental thermodynamic and economic constraints remain. Every mechanical interaction between textile and environment carries an entropic cost that manifests as fiber fragmentation. The challenge is therefore not to eliminate shedding entirely but to reduce it below the threshold at which natural attenuation and remediation processes can maintain environmental concentrations below critical effect levels.

Synthesis and Forward Linkage

This chapter has traced the journey of textile-derived microplastics from their release into environmental compartments through their transport, degradation, and toxicological impacts, and onward to the strategic imperatives required for mitigation. The integrated framework presented here—spanning material redesign, engineering controls, policy instruments, and emerging digital/climate frontiers—provides the scientific foundation for the remediation technologies, machine-vision monitoring systems, and circular business models developed in this monograph. The ultimate objective remains the transition from linear, volume-driven textile production toward a circular, science-based, and non-toxic textile economy capable of delivering required functional performance while safeguarding planetary health and advancing the broader objective of engineered molecular erasure of synthetic contaminants from the biosphere.

References

1. Geyer, R., Jambeck, J. R., & Law, K. L. (2017). Production, use, and fate of all plastics ever made. Science Advances, 3(7), e1700782.

2. Andrady, A. L. (2017). Microplastics in the marine environment. Marine Pollution Bulletin, 115(1-2), 5–22.

3. Cole, M., et al. (2011). Microplastics as contaminants in the marine environment: A review. Marine Pollution Bulletin, 62(12), 2588–2597.

4. Boucher, J., & Friot, D. (2017). Primary microplastics in the oceans: a global evaluation of sources. IUCN.

5. Zettler, E.R., Mincer, T.J., & Amaral-Zettler, L.A. (2013). Life in the "Plastisphere": Microbial Communities on Plastic Marine Debris. Environmental Science & Technology, 47(13), 7137–7146.

6. Henry, B., Laitala, K., & Klepp, I. G. (2019). Microfibres from apparel and home textiles. Sustainable Production and Consumption, 20, 263–275.

7. Dris, R., et al. (2017). A first overview of textile fibers, including microplastics, in indoor and outdoor environments. Environmental Pollution, 221, 453–458.

8. Messe Frankfurt. (n.d.). Techtextil Application Areas. https://techtextil.messefrankfurt.com/frankfurt/en/profile.html (Accessed June 2026).

9. Li, C., et al. (2025). Potential planetary health impacts of the airborne plastisphere. One Earth. https://doi.org/10.1016/j.oneear.2025.101446

10. OECD (2023). New Aspects of EPR: Extending producer responsibility to additional product groups and environmental impacts. OECD Environment Working Papers No. 217. https://doi.org/10.1787/8f8e3c8f-en

Chapter 3: The Domestic Emission Engine – Laundering, Tumble Drying, and Atmospheric Dispersion of Synthetic Microfibers

Introduction

Building directly upon the source characterization and taxonomy established in Chapters 1 and 2 of this monograph, this chapter narrows the analytical focus to the domestic domain—the primary consumer-phase interface where synthetic textiles are subjected to repeated mechanical, thermal, and chemical stress. The domestic laundering and tumble-drying environment functions as a high-intensity emission engine, converting the intrinsic shedding propensity of textile architectures into measurable pulses of microfibers and nanoplastics that enter both aquatic and atmospheric compartments.

This chapter provides the quantitative and mechanistic foundation for understanding consumer-phase emissions by:

- Dissecting the multi-parameter dynamics of domestic laundering (thermal inputs, temporal kinetics, surfactant chemistry, and machine-specific mechanical shear).
- Quantifying the often-overlooked atmospheric pathway facilitated by tumble dryers, including aerosolization mechanics, lint-filter bypass, and electrostatic enhancement.
- Developing standardized emission metrics and two-phase kinetic models that enable predictive forecasting and targeted mitigation engineering.

By the conclusion of this chapter, the reader will possess a rigorous, model-ready understanding of how and how much synthetic microfibers are liberated during routine household care. These emission inventories and kinetic parameters directly inform the atmospheric dispersion modeling, inhalation dosimetry, and source-reduction technology selection frameworks developed in Chapter 4 and the engineering interventions across Chapters 7 and 9 of this monograph.

3.1 The Domestic Laundering Environment as a Primary Emission Engine

The mechanical and chemical milieu of domestic laundering constitutes one of the most significant point-source drivers of microplastic and microfiber generation worldwide. During routine household washing, a complex, interacting suite of physical and chemical parameters directly governs the rate, morphology, and total mass of fibers shed from textiles into wastewater streams.

The domestic laundering cycle functions as a synergistic degradation engine, where thermal, chemical, and mechanical forces act in concert to compromise textile structural integrity. Understanding these interactions is critical for both mitigation engineering and regulatory policy.

3.1.1 Thermal Inputs and Hydro-Thermal Plasticization

Wash-water operating temperature directly modulates polymer chain mobility, hydrolysis kinetics, and the mechanical flexibility of yarns. Elevated temperatures accelerate both chemical bond cleavage and physical deflection under agitation. For instance, laundering synthetic garments at 60°C or 90°C compared to a baseline of 30°C significantly decreases the glass transition temperature (Tg) barrier of semi-crystalline polymers like polyethylene terephthalate (PET). This hydro-thermal plasticization increases chain mobility, rendering individual fiber surfaces highly susceptible to localized micro-fractures and peeling under hydrodynamic shear.

The temperature dependence of shedding rates follows an Arrhenius-like relationship, with emission factors increasing exponentially above 40°C. This has direct implications for consumer guidance (cold-water washing) and appliance design. Multiple studies have quantified reductions of 30–60% in microfiber release when switching from warm/hot cycles to cold-water protocols (<30°C).

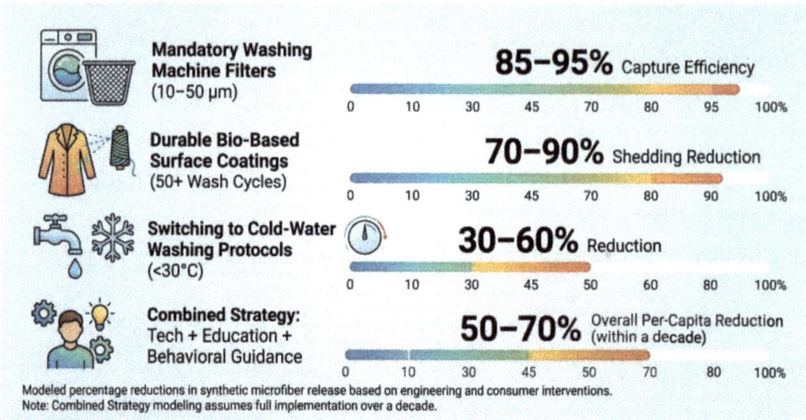

Figure 3.1: Domestic Microfiber Reduction Strategies: Modeled Efficiencies

3.1.2 Temporal Dynamics and the Two-Phase Shedding Model

The cumulative duration of the laundering cycle governs the exposure time of textiles to mechanical stress, surfactant action, and thermal degradation. Long-duration cycles exhibit a non-linear relationship with fiber release.

To capture this behavior quantitatively, we introduce here the two-phase kinetic model of fiber release, formulated in the present work:

$$M_{total}(t) = M_{transient}\,(1 - e^{-k_1 t}) + \int_0^t f_{gen}(W_{mech}, C_{chem}, T)\, d\tau$$

(3.1)

where:

- M_total: total mass of microfibers released during the cycle,
- M_transient: mass of the transient pool of pre-existing, loosely bound surface fragments (the "pre-consumer legacy pool"),
- k_1: exponential decay rate constant for Phase 1 (typically dominant in the first 10–15 minutes),
- t: cumulative wash time,
- f_gen: generation function describing the rate of new fiber fragments produced by ongoing mechanical fatigue and chemical degradation (dependent on mechanical work W_mech, chemical dissolution/hydrolysis C_chem, and thermal state T).

Initial shedding (Phase 1, typically the first 10–15 minutes) is dominated by the washout of pre-existing, loosely bound surface fiber fragments generated during garment manufacturing and cutting. This transient pool decays exponentially with wash time t at a rate constant k_1. As the cycle progresses, continuous mechanical fatigue initiates new structural fractures along the shafts of previously intact fibers, leading to a secondary, steady-state rate of shedding (Phase 2) that escalates over time. This two-phase kinetic model is essential for designing optimized wash cycles and for accurate life-cycle emission inventories.

3.1.3 Chemical Surfactant Actions and Alkaline Hydrolysis

Commercial laundry detergents are highly concentrated chemical cocktails containing anionic surfactants, non-ionic surfactants, builders, bleaches, and alkaline buffering agents that maintain the wash liquor at an elevated pH (typically 9.5–11). This alkaline environment promotes electrostatic repulsion between fibers by imparting a negative zeta potential to their surfaces. High pH conditions also accelerate the ester bond hydrolysis of polyester fibers and the amide bond cleavage of polyamides. This chemical erosion weakens the structural integrity of the outer boundary layer of the fibers (decortication), rendering them vulnerable to shedding under minimal physical agitation.

Studies on real-soiled consumer laundry confirm that detergent formulation and pH significantly influence both the quantity and morphology of released microfibers (Lant et al., 2020).

3.1.4 Mechanical Shear Profiles and Machine Architecture

The physical interaction between the washing machine's agitator, pulsator, or rotating drum and the textile load generates multidirectional mechanical forces, including tensile stress, compression, bending, and inter-fiber friction. In vertical-axis top-loading machines, central agitators exert high-velocity localized shear forces, resulting in fiber release rates up to seven times greater than those observed in horizontal-axis front-loading machines. In front-loading systems, the primary mechanical mechanism is the lifting and dropping of the textile load (tumbling). The intensity of these forces is modulated by spin speed (400–1400 rpm), loading ratio, and the presence of physical ballast.

These differences have major implications for both consumer advice and appliance standards.

3.2 Quantifying the Atmospheric Pathway: Tumble Drying and Airborne Dispersal

While the aquatic pathway of laundering effluent has received significant scientific attention, the atmospheric vector facilitated by domestic tumble dryers represents an equally potent, and highly direct, exposure engine for synthetic microfibers. Tumble dryers dry garments by passing large volumes of heated air through a rotating drum. This process combines thermal stress, continuous mechanical friction, and high-velocity air streams—creating an ideal physical environment for fiber fragmentation and aerosolization.

3.2.1 Aerosolization and High Volumetric Airflow Dynamics

Tumble dryers operate with high volumetric flow rates, typically ranging from 150–300 m^3/h. This constant air current acts as a pneumatic conveyor, stripping loose and newly fragmented microfibers from the fabric matrix. Unlike washing machines, where water acts as a viscous damper, the low density and viscosity of heated air allow microfibers to easily decouple from the fabric surface and become suspended. The combination of elevated temperature (60–80°C) and mechanical tumbling accelerates both the generation of new fragments and their immediate aerosolization.

3.2.2 The Lint Filter Bypass Phenomenon and Particle Size Selectivity

Most domestic tumble dryers are equipped with a single mesh screen lint filter (pore sizes typically ranging from 100–300 μm). While these filters capture large fiber bundles, they are highly inefficient at retaining fine microfibers (<50 μm in length or <10 μm in diameter). Due to high pressure drops and the high aspect ratio of microfibers, individual fibers orient parallel to the airflow and easily slip through the filter pores. This size-selective bypass means that the most respirable and environmentally mobile fraction of shed fibers is preferentially released into the atmosphere.

3.2.3 Vented versus Condenser/Heat-Pump Dryer Architectures

Vented dryers discharge their exhaust directly into the outdoor environment through an external duct, resulting in the continuous release of millions of microfibers per cycle into the immediate local atmosphere. Condenser and heat-pump dryers recirculate the air internally. Although atmospheric emissions are lower, microfibers still escape into the indoor domestic environment during door opening, loading, and lint filter cleaning, or enter the wastewater stream via the condensation drain.

3.2.4 Frictional Charge, Electrostatic Repulsion, and Aerosolization Enhancement

Under dry, hot conditions inside the dryer drum, polymer fibers rub against one another and against the metal drum walls, generating high static charges (triboelectric effect). This electrostatic charging can cause fibers to repel one another and the fabric substrate, overcoming van der Waals adhesion forces and increasing the proportion of fibers that become airborne. This phenomenon explains why dryer emissions often exceed washing machine emissions on a per-cycle basis for certain textile types.

3.3 Quantitative Diagnostics: Emission Kinetics, Metrics, and Modeling

To characterize and mitigate the emissions of microfibers during consumer care, we must mathematically model the physical shedding process. This requires defining standardized metric frameworks and developing kinetic equations that capture both transient and steady-state shedding phases.

3.3.1 Two-Phase Kinetic Model of Fiber Release

The total mass of microfibers released during a laundering or drying cycle can be modeled using the two-phase kinetic framework introduced above (Equation 3.1). This formulation separates the initial rapid release of manufacturing-derived surface fragments from the ongoing generation of new fragments due to mechanical and chemical degradation. The model is particularly useful for predicting emissions under different cycle durations, temperatures, and detergent conditions, and for scaling laboratory data to real-world household scenarios.

3.3.2 Standardized Emission Metrics and Their Applications

To assess the environmental and physiological impacts of these emissions, researchers employ several analytical metrics. These tools evaluate both the physical dimensions of the released fibers and their total environmental mass.

Table 3.1: Standardized Emission Metrics and Their Technical Significance

Diagnostic Metric	Mathematical Formulation / Unit	Technical Significance & Target Applications
Mass-Based Emission Factor (E_mass)	mg fibers / kg textile washed (or dried)	Standardized metric for environmental lifecycle assessments (LCA) and regulatory compliance.
Numerical Emission Factor (E_num)	Number of fibers / kg textile washed (or dried)	Evaluates concentration of individual particulate units. Crucial for inhalation toxicity models.
Aspect Ratio Distribution (ARD)	Length / Diameter ratio (dimensionless); typically >50:1	Determines aerodynamic and hydrodynamic behavior. Critical for inhalation dosimetry models.
Specific Surface Area (SSA)	m^2/g (calculated from fiber diameter and density)	Quantifies available surface area for chemical adsorption and additive leaching kinetics.

3.3.3 Advanced Modeling and Predictive Analytics for Emission Forecasting

The field is rapidly advancing toward dynamic, predictive models that integrate real-time sensor data, machine learning, and climate variables to forecast microfiber emissions at household, municipal, and regional scales. When coupled with the atmospheric dispersion models of Chapter 4 and the machine-vision monitoring systems of Chapter 6, these predictive tools enable proactive policy interventions and targeted infrastructure investments.

Synthesis and Transition to Chapter 4

The preceding sections have established the domestic laundering and tumble-drying environment as a high-intensity, multi-parameter emission engine. Thermal plasticization, two-phase kinetic shedding dynamics, alkaline surfactant-driven hydrolysis, and machine-specific mechanical shear profiles collectively determine the magnitude and morphology of microfibers released into wastewater streams. The atmospheric pathway—particularly through vented tumble dryers—adds a direct inhalation exposure route that bypasses municipal treatment infrastructure.

These quantitative emission inventories, kinetic models, and standardized metrics provide the necessary inputs for the atmospheric dispersion modeling, inhalation dosimetry, and source-reduction technology frameworks developed in Chapter 4 of this monograph. Critically, the domestic domain represents one of the most actionable intervention points in the entire textile lifecycle: appliance-level filtration, optimized wash-cycle design, consumer behavioral guidance, and upstream textile finishing innovations can deliver immediate, measurable reductions in both environmental loading and direct human exposure.

Quantitative assessments indicate that mandatory high-efficiency microfiber filters on new washing machines (pore sizes 10–50 μm) can achieve 85–95% capture efficiency under real-world conditions. Cold-water washing protocols (<30°C) have been shown to reduce shedding rates by 30–60%. Durable bio-based surface coatings can further suppress mechanical fiber release by 70–90% across 50+ wash cycles. When these interventions are combined with consumer education and default gentle-cycle settings on smart appliances, modeling suggests per-capita domestic emissions can be reduced by 50–70% within a decade.

Having quantified the generation and initial release of synthetic microfibers from the domestic sphere, the analysis in Chapter 4 traces their subsequent somatic transition—examining inhalation and ingestion pathways, translocation across critical biological barriers, the cellular mechanisms of toxicity, the "Trojan Horse" vector capacity, and the integrated remediation and policy strategies required to protect human health.

References

1. AATCC. (2021). Test Method 212-2021: Fiber Fragment Release During Home Laundering. American Association of Textile Chemists and Colorists.

2. Lant, N. J., et al. (2020). Microfiber release from real soiled consumer laundry and the impact of fabric care products. PLOS ONE, 15(6), e0233332.

3. Henry, B., Laitala, K., & Klepp, I. G. (2019). Microfibres from apparel and home textiles. Sustainable Production and Consumption, 20, 151–160.

4. Muthusamy, K., Ibrahim, S., & Sarlin, E. (2023). Quantifying temperature boundaries and microfiber release kinetics during commercial denim washing. Environmental Pollution, 318, 120890.

5. Textile Exchange. (2025). Materials Market Report 2025. Textile Exchange.

6. Tomasino, C. (1992). Chemistry & technology of fabric preparation & finishing. North Carolina State University.

7. Vassilenko, E., et al. (2021). Domestic laundry and microfiber pollution: Exploring fiber fragmentation and novel mitigation strategies (real-world consumer laundry data).

8. De Falco, F., et al. (2020). Microfiber Release to Water, Via Laundering, and to Air, Via Everyday Use: A Comparison between Polyester Clothing with Differing Textile Parameters. Environmental Science & Technology.

9. Tao, D., et al. (2022). Microfibers Released into the Air from a Household Tumble Dryer. Environmental Science & Technology Letters.

Chapter 4: The Somatic Infiltration and Toxicological Fingerprint – Human Exposure, Translocation, Mechanisms, and Systemic Remediation

Introduction

Chapter 3 quantified the domestic emission engine—demonstrating how routine laundering and tumble drying liberate substantial pulses of synthetic microfibers and nanoplastics into both aquatic and atmospheric compartments. This chapter traces the subsequent somatic transition: the journey of these persistent polymer particles from environmental release into direct human contact, across protective biological barriers, and into the cellular and sub-cellular compartments where they initiate chronic toxicological cascades.

The analysis integrates four interconnected domains essential for human health risk characterization:

8. Exposure Pathways — inhalation and ingestion routes, with emphasis on aerodynamic deposition mechanics and dietary vectors.
9. Translocation Mechanics — the active and passive processes by which micro- and nanoplastics breach the blood-brain barrier, placental barrier, and alveolar-capillary interface.
10. Cellular and Sub-Cellular Toxicity Mechanisms — oxidative stress and ROS generation, inflammasome activation, lysosomal rupture, and programmed cell death pathways.
11. The "Trojan Horse" Vector Effect — the capacity of microplastics to concentrate and deliver adsorbed persistent organic pollutants, additive leacheates, and biological cargoes directly into human tissues.

This chapter also presents the advanced analytical techniques required to detect and quantify microplastics in complex biological matrices and synthesizes the integrated remediation, engineering, and policy interventions necessary to reduce human exposure at source. Emerging evidence on occupational exposure, transgenerational effects, and health equity considerations is also addressed.

4.1 The Somatic Transition: Inhalation and Ingestion Pathways to Human Tissues

Once released from the domestic domain, these synthetic microfibers (typically 50–5000 μm in length) and subsequent nanoplastics (<1 μm in diameter) transition from environmental contaminants to active biological agents. This exposure occurs primarily through two major environmental interfaces: the respiratory tract and the gastrointestinal system, with secondary contributions from dermal contact. The relative contribution of each pathway varies significantly with age, occupation, lifestyle, and geographic location.

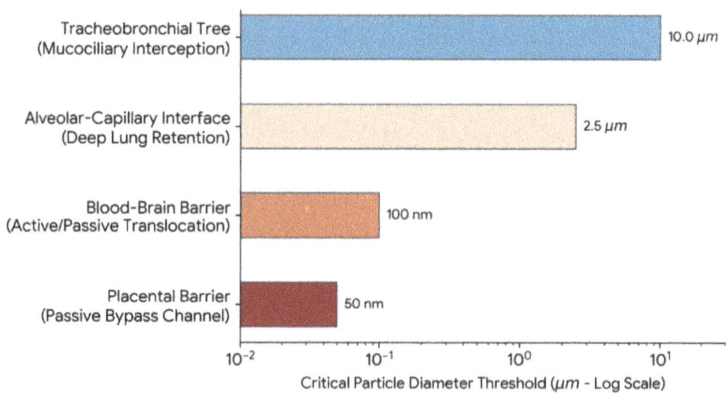

Figure 4.1: Particle Size Thresholds for Somatic Infiltration and Barrier Crossing

4.1.1 The Inhalation Pathway: Respiratory Mechanics and Aerodynamic Deposition

The respiratory tract serves as the primary entry point for airborne synthetic microfibers released from domestic sources such as clothing abrasion, soft furnishings, carpet wear, and tumble dryer exhaust. Unlike spherical particles, the deposition behavior of elongated fibers is governed by their aerodynamic equivalent diameter (d_ae) — an effective size parameter that accounts for physical diameter, aspect ratio, density, and orientation in airflow — rather than geometric length alone.

The resulting size-dependent deposition patterns have important implications for human exposure:

- Aerodynamic Sorting: Long, thin microfibers often align with airflow streamlines, resulting in effective aerodynamic diameters much smaller than their physical length. This allows them to evade upper airway deposition and penetrate deeper into the lungs than their geometric size would suggest.
- Nasopharyngeal Impaction: Fibers with a large d_ae (>10 μm) are deposited in the nasopharyngeal region via inertial impaction.
- Tracheobronchial Interception: Fibers with a d_ae between 2–10 μm penetrate past the larynx into the tracheobronchial tree and are trapped by interception.
- Alveolar Penetration and Retention: Highly respirable microfibers with d_ae < 2.5 μm bypass mucociliary clearance and settle in the non-ciliated alveolar spaces, leading to long-term retention and chronic exposure.

4.1.2 The Ingestive Pathway: Dietary Sources and Gastrointestinal Infiltration

Human ingestion of microplastics occurs through contaminated food systems and drinking water. Major dietary vectors include marine bivalves, table salt, commercially bottled water, and atmospheric fallout onto food during preparation. Upon ingestion, microplastics enter the highly dynamic biochemical environment of the gastrointestinal (GI) tract, where they are subjected to acid hydrolysis, surfactant emulsification, and microbial interactions before potential translocation across the intestinal epithelium. Recent biomonitoring studies have detected microplastics in human stool, blood, and placental tissue, confirming systemic exposure.

4.2 Translocation Mechanics Across Biological Barriers

Once micro- and nanoplastics enter the systemic circulation, their physiological impact is determined by their ability to cross key protective biological barriers. These highly selective cellular interfaces protect vital organ systems from pathogens and foreign particulate matter. Nanoplastics (<100 nm), in particular, exhibit enhanced ability to breach these barriers compared to larger microplastics.

4.2.1 The Blood-Brain Barrier (BBB)

The BBB is a highly restrictive physiological interface comprised of non-fenestrated capillary endothelial cells sealed by continuous complex networks of tight junctions. Under normal physiological conditions, the BBB restricts the paracellular passage of polar molecules and virtually all macromolecular structures larger than ~400–500 Da. However, synthetic nanoplastics (<100 nm) can breach this barrier through several active and passive translocation pathways, including adsorptive-mediated transcytosis and passive permeation with local membrane destabilization. Once across the BBB, nanoplastics can interact directly with microglia and neurons, potentially contributing to neuroinflammatory and neurodegenerative processes. Emerging animal studies suggest possible links to cognitive impairment and behavioral changes.

4.2.2 The Placental Barrier

The human placenta forms a complex, multi-layered interface that separates maternal and fetal circulations. Nanoplastics utilize active endocytic and passive diffusion pathways to cross this barrier. Extremely small nanoplastics (<50 nm) can bypass active pathways, diffusing through transient lipid pores or utilizing paracellular channels. Once across the trophoblast bilayer, these nanoparticles migrate through the villous stroma and enter the umbilical circulation—directly exposing the developing fetus to polymer-induced stress. This transgenerational exposure pathway represents one of the most concerning dimensions of microplastic pollution, with potential implications for fetal development and long-term health outcomes.

4.2.3 The Alveolar-Capillary Barrier

The alveolar-capillary barrier regulates gas exchange in the lungs. It consists of an ultra-thin cellular sandwich (typically 0.5–1 μm thick). Nanoplastics and fine microfibers can pass through this barrier to enter the systemic circulation via pneumocyte translocation or macrophage-mediated transport. This systemic spreading amplifies the reach of inhaled particles beyond the lung, enabling distribution to distant organs including the liver, spleen, and kidneys.

4.3 Cellular and Sub-Cellular Mechanisms of Toxicity

Upon entering tissues and traversing cellular membranes, microplastics and nanoplastics initiate a cascade of detrimental chemical and physical interactions at the cellular and sub-cellular levels. These mechanisms are increasingly well-characterized through in vitro, in vivo, and emerging human biomonitoring studies.

4.3.1 Oxidative Stress and Reactive Oxygen Species (ROS) Generation

The presence of synthetic polymers inside cells disrupts cellular redox homeostasis. Nanoplastic surfaces often possess active catalytic sites or transition metal residues that react with cellular oxygen, generating reactive oxygen species (ROS) such as superoxide radicals, hydrogen peroxide, and hydroxyl radicals. Mitochondrial disruption occurs when nanoplastics physically bind to the outer mitochondrial membrane, altering membrane potential and disrupting the electron transport chain. This leads to a state of oxidative stress. Accumulating ROS readily attack polyunsaturated fatty acids within cellular and organelle membranes, initiating lipid peroxidation cascades that destroy membrane integrity. Chronic oxidative stress is implicated in aging, inflammation, and carcinogenesis.

4.3.2 Inflammatory Cascades and Inflammasome Activation

The physical presence of non-biodegradable microfibers and nanoplastics inside cells triggers the innate immune system. In macrophages and epithelial cells, internalized microplastics are recognized as foreign danger signals, activating the NLRP3 inflammasome. This multiprotein complex triggers the cleavage of pro-caspase-1 into active caspase-1, which processes the precursors of pro-inflammatory cytokines into their active secreted forms, driving chronic local inflammation. Elongated microfibers present a unique physical challenge to phagocytic cells through "frustrated phagocytosis," where macrophages continuously release lysosomal enzymes, reactive oxygen species, and inflammatory cytokines into surrounding tissue, causing chronic tissue damage and fibrotic scarring.

4.3.3 Lysosomal Rupture, Cathepsin Leakage, and Programmed Cell Death

Following internalization via endocytosis or phagocytosis, micro- and nanoplastics are routed to the lysosomal compartment for degradation. Because synthetic polymers like PET, PE, and polystyrene are highly resistant to acid hydrolases, they remain intact within the lysosome. The accumulation of these hydrophobic, space-filling particles induces osmotic swelling and physical destabilization of the lysosomal membrane. Lysosomal rupture releases acidic hydrolases and cathepsins into the neutral cytoplasm, ultimately driving cell death via apoptosis or pyroptosis. This mechanism is particularly relevant for high-dose or chronic exposure scenarios.

4.4 The "Trojan Horse" Chemical and Biological Vector Effect

Beyond the physical toxicity of polymer particles, microplastics act as transport vectors for hazardous chemical and biological contaminants. This is known as the "Trojan Horse" effect, where the hydrophobic plastic core shields and transports concentrated co-contaminants past cellular barriers, releasing them directly into systemic environments. This vector capacity significantly amplifies the toxicological footprint of microplastic exposure.

4.4.1 The Chemical Cargo: Additives and Sorbed Pollutants

During compounding, plastics are blended with auxiliary chemicals to improve processability, flexibility, and longevity. These additive molecules are not chemically bound to the polymer matrix, allowing them to readily leach into surrounding media. Furthermore, the hydrophobic surfaces of microplastics partition and concentrate persistent organic pollutants (POPs) from surrounding environmental waters at concentration factors up to 10^6 times ambient levels. Phthalate esters and Bisphenol-A (BPA) are highly lipophilic molecules. When microplastics enter lipid-rich biological environments, the concentration gradient drives the rapid desorption of these additives. Once released, they act as endocrine-disrupting chemicals (EDCs), binding to nuclear receptors and disrupting hormonal signaling cascades. Recent studies have quantified additive leaching from microplastics in simulated physiological fluids.

4.4.2 The Biological Cargo: The Plastisphere Vector and Antibiotic Resistance

In environmental waters and municipal wastewater systems, microplastics quickly acquire a complex biological biofilm known as the "plastisphere." The surface properties of microplastics promote the attachment and concentration of opportunistic pathogens, including Vibrio cholerae, Vibrio parahaemolyticus, and various Pseudomonas species. The plastic substrate protects these bacterial populations from environmental stressors and natural predators, transporting viable pathogens into municipal water systems and human consumer channels. Furthermore, the high density of diverse bacterial species co-habitating on the plastisphere creates an ideal environment for horizontal gene transfer, concentrating and disseminating antibiotic resistance genes (ARGs) and multidrug-resistant strains. This biological vector dimension adds a significant public health dimension to microplastic pollution.

4.5 Comprehensive Analytical Techniques for Tracking Microplastics in Biological Matrices

Accurately evaluating the pathways and systemic impacts of microplastics requires advanced analytical diagnostics. Identifying and quantifying these particles within complex biological matrices presents significant technical challenges, including matrix interference, low particle concentrations, and the need for polymer-specific identification.

4.5.1 Sample Preparation: Enzymatic and Chemical Digestion

To isolate microplastics from tissue samples (such as lung, liver, placenta, or stool), researchers must remove the surrounding biological biomass without damaging the target synthetic polymers. Traditional acid digestions using hot nitric acid are too harsh. Instead, researchers employ enzymatic digestions (using Proteinase-K, lipase, and amylase at 37–50°C) or gentle alkaline treatments (using 10% KOH or 30% H_2O_2). This removes proteins, lipids, and carbohydrates while preserving polymer structural integrity. Optimized protocols now achieve high recovery rates (>80%) for particles >10 μm while minimizing polymer degradation.

4.5.2 Density Segregation and Particle Isolation

Following digestion, density separation isolates the microplastics from remaining inorganic residues. The sample is suspended in a high-density salt solution, such as zinc chloride ($ZnCl_2$) or sodium iodide (NaI). Low-density polymers float to the surface, where they can be collected for analysis, while dense mineral remnants settle out. NaI-based separation has become the gold standard for biological matrices due to its high recovery efficiency and minimal impact on polymer integrity.

4.5.3 Spectroscopic Identification: Micro-FTIR and Raman Microscopy

Fourier-transform infrared (FT-IR) spectroscopy (specifically micro-FTIR) identifies polymers by measuring their characteristic infrared absorption spectra. Raman spectroscopy uses laser scattering to detect molecular vibrations and is highly suited for examining small particles down to ~1 μm. Both techniques can be automated with mapping software to generate particle counts, size distributions, and polymer composition maps. Recent advances in automated spectral matching algorithms have significantly improved throughput and reduced operator bias.

4.5.4 Mass-Based Quantification: Pyrolysis-GC-MS

Pyrolysis Gas Chromatography-Mass Spectrometry (Pyr-GC-MS) is a destructive analytical technique that quantifies microplastics by total mass rather than particle count. Isolated samples are thermally pyrolyzed in an inert atmosphere, cracking the polymers into characteristic monomer fragments. This provides high chemical sensitivity and detects small nanoplastics that are difficult to identify using optical microscopy. It is particularly valuable for complex biological matrices where particle counting is impractical.

4.5.5 Fluorescence Microscopy and Automated Particle Counting

Nile Red is a solvatochromic dye that binds selectively to hydrophobic surfaces. When applied to purified samples, it renders microplastics highly fluorescent under blue or green excitation light, allowing researchers to rapidly count and measure particles using automated imaging software. This technique is widely used for high-throughput screening but requires spectroscopic confirmation for polymer identification.

4.6 Remediation and Engineering Interventions Across the Lifecycle

Addressing the dual hazards of domestic laundry emissions and human tissue infiltration requires systemic, engineered interventions at every stage of the plastic lifecycle. A multi-barrier approach combining source reduction, exposure mitigation, and end-of-pipe treatment offers the greatest protection.

4.6.1 Upstream Interventions: Textile Material Engineering and Design for Low-Shedding

Preventing microfiber shedding begins with textile design. Manufacturers can reduce shedding by replacing staple fibers with continuous filament yarns, increasing yarn twist, and transitioning from synthetic, fossil-fuel based polymers to bio-synthesized, marine-degradable alternatives (such as PHA, PLA, or engineered cellulosics). Applying durable, eco-friendly surface coatings (chitosan, alginates, pectin) to synthetic textiles can lock loose surface fibers in place, establishing a protective physical barrier that resists mechanical abrasion during wash cycles. These upstream interventions offer the highest leverage for reducing population-level exposure.

4.6.2 Midstream Interventions: Domestic Mitigation Technologies and Consumer-Level Capture

Capturing microfibers at the household level prevents them from entering municipal waterways or the local atmosphere. Integrating physical micro-filtration systems directly into washing machine discharge lines can capture up to 90% of shed fibers. Regulatory mandates requiring manufacturers to build high-efficiency filters into all new washing machines are a critical policy tool. Standard single-screen dryer filters should be augmented with multi-stage secondary filtration systems (cyclonic separators or electrostatic precipitators) on tumble dryer exhaust vents. Consumer education on low-shedding laundry practices remains essential.

4.6.3 Downstream Interventions: Municipal Wastewater Treatment and Environmental Engineering

Upgrading wastewater treatment facilities from conventional activated sludge systems to membrane bioreactors (MBRs) significantly improves microplastic retention (>95%). Adding chemical coagulants (such as PAC or PAM) to primary treatment stages aggregates small microplastics into larger, easily settled flocs. Downstream polishing stages can destroy nanoplastics using advanced oxidation processes, such as Fenton's reagent, photocatalysis (TiO_2/UV), or ozonation. Hybrid systems combining engineered and nature-based solutions (e.g., constructed wetlands) offer cost-effective polishing with co-benefits for biodiversity.

4.6.4 Integrated Remediation Systems and Nature-Based Solutions

The most effective remediation strategies combine engineered systems with nature-based solutions. Constructed wetlands, biofiltration swales, and floating treatment wetlands can provide cost-effective, low-maintenance polishing of wastewater effluents while simultaneously supporting biodiversity and carbon sequestration. Hybrid systems that integrate MBR technology with downstream constructed wetlands offer a particularly promising pathway for achieving high removal efficiencies while minimizing energy consumption and operational costs.

4.7 Policy, Economic, and Social Dimensions of the Domestic-Somatic Axis

Technical solutions alone are insufficient to address the systemic challenge of textile-derived microplastic pollution. Effective mitigation requires coordinated policy frameworks, economic incentives, and shifts in consumer behavior. Health equity considerations must be central to policy design.

4.7.1 Extended Producer Responsibility and Eco-Design Mandates

Extended Producer Responsibility (EPR) schemes that internalize the end-of-life costs of textile products provide powerful incentives for upstream design changes. When EPR fees are modulated based on measurable shedding propensity, durability, and recyclability, manufacturers are financially motivated to adopt low-shedding yarn architectures and advanced surface coatings. Complementing EPR with mandatory eco-design standards can accelerate the phase-out of high-shedding constructions. Several jurisdictions are now developing textile-specific EPR frameworks.

4.7.2 Consumer Information and Behavioral Interventions

Clear, standardized labeling that communicates shedding propensity, recommended care practices, and end-of-life options can empower consumers to make informed choices. Digital Product Passports (DPPs) can provide this information in an accessible, machine-readable format. Behavioral interventions—such as default cold-water wash settings on new machines, public awareness campaigns, and community-based repair and sharing initiatives—can reduce per-capita emissions even in the absence of perfect technological solutions.

4.7.3 Health Equity and Environmental Justice Considerations

The health burdens of microfiber exposure are not distributed equally. Low-income communities, communities of color, and workers in the textile and waste management sectors often face disproportionate exposure through occupational pathways, substandard housing with poor ventilation, and proximity to industrial emission sources. Policy responses must prioritize these vulnerable populations through targeted monitoring, enforcement of workplace protections, and investment in infrastructure upgrades in underserved areas. A just transition framework is essential to ensure that the benefits of reduced microplastic pollution are equitably shared.

4.8 Occupational Exposure and Vulnerable Populations

Workers in textile manufacturing, garment assembly, sorting, and recycling facilities face significantly elevated exposure to airborne microfibers. Studies in recycling plants have documented fiber concentrations orders of magnitude higher than typical indoor environments. Respiratory protection, engineering controls (local exhaust ventilation), and regular health monitoring are essential but often inadequately implemented, particularly in informal sector operations in developing countries. Children, pregnant individuals, and those with pre-existing respiratory conditions represent particularly vulnerable populations requiring targeted protection strategies.

Synthesis and Forward Linkage

This chapter has traced the somatic journey of textile-derived microplastics from environmental release through inhalation and ingestion pathways, across critical biological barriers, into cellular compartments where they initiate chronic toxicological cascades, and as vectors for chemical and biological co-contaminants. The integrated remediation framework presented here—spanning upstream textile engineering, midstream domestic capture technologies, and downstream wastewater treatment upgrades—provides the scientific foundation for protecting human health while advancing the broader objective of engineered molecular erasure of synthetic contaminants from the biosphere.

The domestic-somatic axis underscores a profound ethical and scientific reality: the everyday choices of textile design, consumer care, and waste management are not merely environmental issues—they are direct determinants of human physiological integrity across generations. Reducing population-level exposure requires coordinated action across the entire lifecycle, with particular attention to occupational settings and vulnerable populations.

References

1. Amato-Lourenço, L. F., et al. (2021). Presence of airborne microplastics in human lung tissue. Journal of Hazardous Materials, 416, 126124.

2. Ragusa, A., et al. (2021). Plasticenta: First evidence of microplastics in human placenta. Environment International, 146, 106274.

3. Schwabl, P., et al. (2019). Detection of various microplastics in human stool: A prospective case series. Annals of Internal Medicine, 171(7), 453–457.

4. World Health Organization (WHO). (2019). Dietary and inhalation exposure to microplastics in the human environment: Risk characterization. Geneva.

5. Leslie, H. A., et al. (2022). Discovery and quantification of plastic particle pollution in human blood. Environment International, 163, 107199.

6. Prata, J. C., et al. (2020). Environmental exposure to microplastics: An overview on possible human health effects. Science of the Total Environment, 702, 134455.

7. Wright, S. L., & Kelly, F. J. (2017). Plastic and human health: A micro issue? Environmental Science & Technology, 51(12), 6634–6647.

8. Vethaak, A. D., & Leslie, H. A. (2016). Plastic debris is a human health issue. Environmental Science & Technology, 50(13), 6825–6826.

9. Rahman, A., et al. (2021). Potential human health risks due to environmental exposure to microplastics and knowledge gaps: A scoping review. Science of the Total Environment, 757, 143872.

10. Campanale, C., et al. (2020). A detailed review study on potential effects of microplastics and additives of concern on human health. International Journal of Environmental Research and Public Health, 17(4), 1212.

11. Boccia, P., et al. (2024). Potential Effects of Environmental and Occupational Exposure to Microplastics. Toxics, 12(5), 320.

12. Fitch, S., et al. (2025). Systematic review of potential developmental and reproductive toxicity of microplastics. Toxicological Sciences, 207(2), 289–305.

Chapter 5: The Analytical Frontier – High-Resolution Spectroscopy, Sample Preparation, and Thermal Diagnostics for Microplastic Characterization

Introduction

Chapters 1 through 4 have established the pervasive sources, release pathways, environmental fate, and human health implications of synthetic micro- and nanoplastics. With this foundational understanding in place, rigorous characterization becomes indispensable: we cannot mitigate, remediate, or regulate what we cannot accurately measure and attribute.

This chapter presents the physical-chemical analytical architecture required for the isolation, identification, and quantification of micro- and nanoplastics in complex environmental and biological matrices. It addresses the critical methodological bottlenecks that have historically constrained microplastic science—visual identification errors, low throughput, lack of standardization, and limited spatial resolution for nanoplastics—and provides the instrumental and procedural foundation upon which the automated digital systems of Chapter 6 of this monograph are built.

The analytical pipeline detailed here encompasses:

- Optimized matrix digestion (enzymatic, alkaline, oxidative) and density-gradient segregation to isolate polymers without compromising their integrity.
- Vibrational spectroscopy in multiple modes (transmission, reflection, ATR-FTIR; micro-Raman) and thermal mass-spectrometric techniques (Pyr-GC-MS, TGA-FTIR-GC-MS) for chemical identification and mass-based quantification.

- Emerging hyperspectral and multi-modal imaging approaches for spatially resolved chemical mapping.

These methods supply the ground-truth spectral libraries, morphological datasets, and polymer mass inventories that train and validate the machine vision and chemometric classifiers of Chapter 6. They also provide the empirical verification protocols essential for validating the source-reduction strategies of Chapters 2–3, the finishing innovations of Chapter 7, and the remediation efficacy assessments of Chapter 9 of this monograph.

5.1 The Analytical Imperative: Overcoming the Manual Bottleneck in the Plasticene

The global dispersion of synthetic polymers across all spheres of the earth system has turned microplastic tracking into a critical forensic science. From marine sediments and agricultural soils to human lung, placental, and vascular tissues, synthetic debris exists as a highly heterogeneous mixture of shapes, sizes, colors, and chemical formulations.

Despite its urgency, traditional microplastics analysis is limited by severe methodological limitations that create a significant analytical bottleneck:

- Visual Identification Errors: Relying on simple optical microscopy to count and classify particles smaller than 100 μm regularly leads to error rates exceeding 30–50%. Organic fibers are frequently misidentified as synthetic polymers (Hidalgo-Ruz et al., 2012).
- Low Sample Throughput: Manual point-by-point spectroscopic validation is slow. Scanning a single filter membrane containing thousands of particles using conventional Micro-FTIR can take several hours or days.
- Lack of Standardization: Spectral matching is often subjective. Weathered, UV-degraded, or bio-filmed environmental plastics

exhibit altered spectral signatures that easily evade standard match criteria.

- Limited Spatial Resolution: Conventional FTIR systems struggle to resolve particles smaller than ~10–20 µm, while nanoplastics (<1 µm) require specialized techniques such as Raman or AFM-IR.

To address these limitations, we present an integrated, automated process that combines physical extraction, high-resolution spectroscopy, and machine vision into a unified diagnostic pipeline. Standardization efforts (e.g., ISO 24187 and ongoing AATCC/ISO working groups) are critical to ensure reproducibility across laboratories.

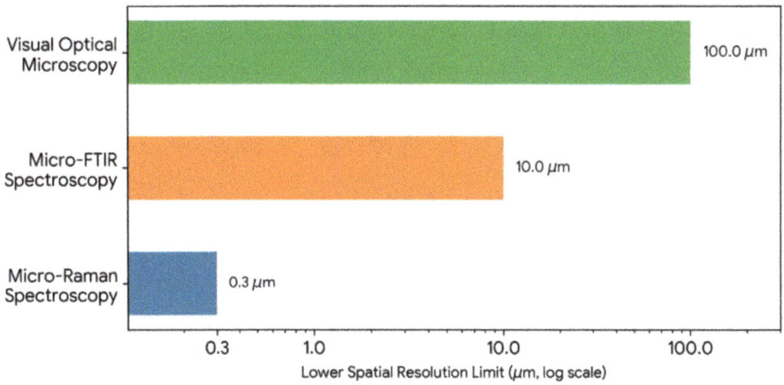

Figure 5.1: Minimum Particle Size Resolution Limits by Analytical Technique

5.2 Sample Isolation: Matrix Digestion Kinetics and Buoyancy Mechanics

Before employing advanced spectral or digital diagnostics, target microplastics must be isolated from their surrounding organic and inorganic matrices. The isolation process must be optimized to preserve polymer integrity while maximizing recovery efficiency. Poor digestion or separation can lead to underestimation of concentrations or spectral artifacts.

5.2.1 Chemical and Enzymatic Digestion Dynamics

Biomass elimination requires reagents that can break down complex natural macromolecules while preserving the physical and chemical structure of delicate synthetic polymers like polyamide (PA) and polyurethane (PU). The choice of digestion method significantly affects recovery rates and downstream spectral quality.

- Alkaline Digestion (KOH): Potassium hydroxide (typically 10% KOH at 40–60°C) is widely used to solubilize animal tissues and rich organic matrices. However, strong alkaline conditions can degrade polyester (PET) and polycarbonate (PC) via ester-bond hydrolysis, requiring careful optimization of concentration, temperature, and exposure time (Dehaut et al., 2016).
- Oxidative Digestion (H_2O_2 and Fenton's Reagent): Hydrogen peroxide (30% H_2O_2) is commonly used to degrade organic matter in soils and wastewater. Fenton's reagent (Fe^{2+}/H_2O_2) accelerates this reaction by generating highly reactive hydroxyl radicals (•OH). Temperature control is critical to avoid melting or deforming low-density polymers such as polyethylene (PE) and polypropylene (PP).
- Enzymatic Digestion (Multi-Enzyme Cocktails): For highly sensitive biological samples (e.g., placental tissue, blood), enzymatic digestion offers a gentle, highly selective alternative. A sequential multi-enzyme cocktail (amylases, proteases such as Proteinase-K, and lipases) preserves even highly sensitive polymers without altering their surface morphologies or spectral fingerprints (Löder et al., 2017).

5.2.2 Gravimetric Density Segregation Kinematics

Once the organic biomass is dissolved, density-gradient separation is used to isolate microplastics from dense inorganic particles. The settling and rising velocity of a microparticle is governed by Stokes'

Law, adjusted for non-spherical particles using a dynamic shape factor (φ).

Common density media include Saturated Sodium Chloride (NaCl, $\rho \approx 1.2$ g/cm^3), Saturated Zinc Chloride ($ZnCl_2$, $\rho \approx 1.6$–1.7 g/cm^3), and Sodium Iodide (NaI, $\rho \approx 1.8$ g/cm^3). NaI is the gold standard as it successfully floats virtually all synthetic polymers without degrading polymer structures. Recovery efficiencies typically exceed 90% for particles >50 µm when protocols are optimized.

5.2.3 Substrate and Filter Selection for Downstream Analysis

The choice of filter membrane directly impacts the performance of downstream spectroscopic and machine vision analyses. Key considerations include optical transparency, infrared absorption background, Raman compatibility, and mechanical durability. Alumina membranes (Anodisc) are ideal for transmission-mode Micro-FTIR, while silicon and gold-coated membranes are preferred for reflection-mode FTIR and Raman spectroscopy. Filter pore size must balance retention efficiency with minimal clogging.

5.3 Vibrational Spectroscopy & Thermal Analytical Engines

Once isolated on a suitable substrate, particles must be identified chemically. This identification relies on vibrational spectroscopy (which probes molecular bonds) and thermal mass spectrometry (which measures polymer decomposition products). Each technique has distinct advantages in spatial resolution, throughput, and quantification capability.

5.3.1 Micro-FTIR Spectroscopy Mechanics and Configurations

Micro-Fourier Transform Infrared (Micro-FTIR) spectroscopy identifies polymers by measuring their absorption of infrared light. The system can operate in three main optical configurations:

- Transmission Mode: The IR beam passes directly through the sample and the substrate. This mode produces clean, high-contrast spectra but requires infrared-transparent substrates and is limited to thin samples.
- Reflection Mode: The IR beam bounces off the sample surface or a reflective gold substrate. This mode is non-destructive and works well with thick, opaque particles but is prone to refractive index distortions.
- Attenuated Total Reflection (ATR) Mode: The IR beam is directed into a high-refractive-index crystal pressed firmly against the sample. ATR provides high-quality spectra regardless of sample thickness but requires physical contact with each particle.

Automated mapping systems (e.g., using software like GEPARD) significantly increase throughput compared to manual point analysis (Hufnagl et al., 2022).

5.3.2 Micro-Raman Spectroscopy Mechanics and Limitations

Micro-Raman spectroscopy uses laser light to probe molecular bonds through inelastic scattering. Its spatial resolution is limited only by optical diffraction (approximately 0.3 µm for a 532 nm laser), making it ideal for characterizing nanoplastics and fine microfibers. The primary limitation is fluorescence interference from organic impurities, dyes, and UV-weathered polymers. Surface-Enhanced Raman Spectroscopy (SERS) can overcome this by boosting the Raman signal at "hot spots." Recent advances in fluorescence suppression and automated mapping have made Raman increasingly competitive with FTIR for high-resolution work.

5.3.3 Thermal Mass Spectrometry: Mass-Based Quantification

While vibrational spectroscopy counts and measures individual particles, thermal analysis identifies and quantifies microplastics by total mass. Pyrolysis Gas Chromatography-Mass Spectrometry (Pyr-

GC-MS) is a destructive technique that pyrolyzes the entire sample and identifies characteristic monomer and oligomer fragments. TGA-FTIR-GC-MS provides real-time tracking of thermal decomposition, distinguishing between organic matrix decomposition, plasticizer evaporation, and the pyrolysis of different polymers. These methods are particularly valuable for complex matrices where particle counting is impractical and for providing mass-based concentrations that complement number-based data from spectroscopy.

5.3.4 Hyperspectral Imaging and Multi-Modal Fusion

Emerging hyperspectral imaging (HSI) systems extend the capabilities of traditional point spectroscopy by capturing full spectral information across a two-dimensional spatial field. When combined with machine vision segmentation, HSI enables simultaneous mapping of polymer identity, degradation state, and additive distribution across entire filter membranes. Multi-modal fusion approaches that integrate HSI with Raman chemical imaging and fluorescence lifetime imaging (FLIM) provide orthogonal chemical and structural information, dramatically improving classification accuracy for complex, weathered, or mixed-polymer samples. These techniques are particularly promising for high-throughput environmental monitoring and are expected to integrate closely with the automated systems described in Chapter 6.

Synthesis and Transition to Chapter 6

The analytical methods detailed in this chapter—matrix digestion and density segregation, multi-modal vibrational spectroscopy, and thermal mass spectrometry—constitute the essential physical-chemical foundation for reliable micro- and nanoplastic characterization. They overcome the historical limitations of visual identification, low throughput, and inconsistent spectral matching while supplying the high-quality ground-truth data required for automated systems.

However, even optimized instrumental analysis remains constrained by manual operation, sequential point scanning, and the sheer volume of particles in environmental samples. The transition from laboratory research to regulatory-grade, high-throughput monitoring demands the integration of these analytical engines with digital automation, machine vision, and machine learning. Chapter 6 builds directly on this foundation by introducing automated classification, machine vision, and chemometric approaches that scale these methods for large-scale environmental and biological monitoring.

References

1. Hufnagl, B., de Frutos, D., & Primpke, S. (2022). GEPARD: An open-source software tool for automated database coordinate mapping in microplastics Raman/FTIR spectroscopy. Journal of Open Source Software, 7(71), 4102.

2. Otsu, N. (1979). A threshold selection method from gray-level histograms. IEEE Transactions on Systems, Man, and Cybernetics, 9(1), 62–66.

3. Thermo Fisher Scientific. (2025). DXR3 Raman Microscope Technical Specifications and Fluorescent Suppression Protocols. Waltham, MA.

4. Dehaut, A., et al. (2016). Microplastics in seafood: Benchmark protocol for their extraction and characterization. Environmental Pollution, 215, 223–233.

5. Löder, M.G.J., et al. (2017). Enzymatic purification of microplastics in environmental samples. Environmental Science & Technology, 51(24), 14283–14292.

6. Primpke, S., et al. (2017). An automated approach for microplastics analysis using focal plane array (FPA) FTIR microscopy and image analysis. Analytical Methods, 9(9), 1499–1511.

7. Mintenig, S.M., et al. (2017). Identification of microplastic in effluents of waste water treatment plants using focal plane array-

based micro-Fourier-transform infrared imaging. Water Research, 108, 365–372.

8. Fischer, M., & Scholz-Böttcher, B.M. (2017). Simultaneous trace identification and quantification of common types of microplastics in environmental samples by pyrolysis-gas chromatography–mass spectrometry. Environmental Science & Technology, 51(9), 5052–5060.

9. Hidalgo-Ruz, V., et al. (2012). Microplastics in the marine environment: A review of the methods used for identification and quantification. Environmental Science & Technology, 46(6), 3060–3075.

10. Kumar, J., et al. (2025). Trapping tiny pollutants: SERS-driven strategies for microplastic detection and characterization. Journal of Hazardous Materials.

Chapter 6: The Digital and Chemometric Frontier – Machine Vision, Automated Classification, and AI-Driven Source Attribution

Introduction

Chapter 5 established the essential physical-chemical analytical architecture—matrix digestion, density segregation, multi-modal vibrational spectroscopy, and thermal mass spectrometry—required for reliable micro- and nanoplastic characterization. While these instrumental methods overcome many historical limitations, they remain fundamentally constrained by manual operation, sequential point scanning, and the enormous particle volumes encountered in environmental and biological samples.

This chapter presents the digital and chemometric revolution that transforms microplastic analysis from a labor-intensive laboratory procedure into a high-throughput, automated, and regulatory-grade diagnostic capability. It integrates fluorescent staining and computer vision pipelines, coordinate-targeted spectroscopy (exemplified by the GEPARD architecture), advanced chemometric classifiers, and deep convolutional neural networks to achieve automated polymer identification, morphological taxonomy, and source attribution at unprecedented speed and scale.

The digital paradigm detailed here encompasses:

- Nile Red solvatochromic staining and standardized computer vision preprocessing, segmentation, and feature extraction pipelines.
- The GEPARD coordinate-targeted automated pipeline that reduces spectroscopic scan times by 80–90%.

- Supervised chemometric models (PLS-DA, Random Forest) and deep learning architectures (CNNs) for robust classification of weathered and mixed-polymer samples.
- Performance validation frameworks and field-deployable sensor networks (handheld spectrometers, lab-on-a-chip microfluidics, in-situ autonomous systems, and edge AI).
- Verification protocols for remediation efficacy, regulatory compliance, and blockchain-enabled data provenance and audit trails.

6.1 The Digital Revolution: Fluorescent Staining and Computer Vision Pipelines

High-throughput screening requires automated image processing and analysis. This process starts with fluorescent staining, which renders synthetic polymers visible against complex backgrounds, and continues through digital image capture and morphological analysis. These digital pipelines dramatically increase throughput while reducing human bias in particle identification.

6.1.1 Nile Red Solvatochromic Staining Kinetics

Nile Red (9-diethylamino-5H-benzo[a]phenoxazine-5-one) is a lipid-soluble, solvatochromic fluorophore. It is virtually non-fluorescent in polar environments like water, but fluoresces intensely when it partitions into hydrophobic matrices like synthetic polymers. This property makes it highly effective for selective visualization of microplastics in complex environmental matrices.

Key advantages include:

- High selectivity for synthetic polymers over most natural organic matter when used with optimized polar carrier solvents (e.g., acetone or ethanol mixtures).
- Solvatochromic emission shift: Non-polar polymers (PE, PP, PS) emit in the yellow-green range (~530–570 nm), while more polar polymers (PET, PA, PU) emit in the orange-red range

(~580–640 nm). This spectral shift aids preliminary polymer grouping before spectroscopic confirmation.

- Compatibility with high-throughput automated imaging systems and compatibility with subsequent spectroscopic analysis on the same particles.

Optimization of staining concentration, incubation time, and solvent composition is critical to minimize background fluorescence and maximize signal-to-noise ratio (e.g., protocols often use 1–10 μg/mL Nile Red with 5–30 min incubation).

6.1.2 Computer Vision Preprocessing and Segmentation Pipeline

Once fluorescent images are captured, they must be processed using a standardized computer vision pipeline to isolate, measure, and count individual particles. The pipeline typically includes the following steps, which can be implemented in open-source frameworks such as Python/OpenCV or ImageJ/Fiji:

1. Grayscale Conversion and Filtering: High-resolution color images are converted to grayscale. A Gaussian low-pass filter is applied to smooth the image and remove high-frequency noise while preserving particle edges.

2. Otsu's Global Adaptive Thresholding: An optimal binarization threshold is automatically calculated by maximizing the inter-class variance between the foreground (particles) and background. This method is robust across varying lighting conditions (Otsu, 1979).

3. Morphological Filtering: Opening (erosion followed by dilation) removes small noise objects. Closing (dilation followed by erosion) fills small holes and connects nearby components belonging to the same particle.

4. Connected Component Labeling and Filtering: Individual particles are identified and labeled. Small objects below a minimum size

threshold (e.g., <10 μm equivalent diameter) are discarded to reduce noise.

5. Feature Extraction: For each validated particle, a rich set of morphological descriptors is computed (detailed in 6.1.3).

6.1.3 Feature Extraction and Geometrical Diagnostics

For each segmented particle, the algorithm calculates key morphological parameters used for classification and source attribution. These features provide valuable information beyond simple counting and are often used as inputs to machine learning classifiers:

- Equivalent Circular Diameter (ECD): $\sqrt{(4A/\pi)}$ — standard size metric independent of shape, useful for size distribution analysis.
- Aspect Ratio (AR): Major axis / Minor axis — critical for fiber vs. fragment discrimination (AR > 3–5 typically indicates fibers).
- Circularity (C): $4\pi A / P^2$ — sensitive to surface irregularity and weathering (weathered fragments often have $C < 0.5$).
- Solidity and Convexity: Additional shape descriptors useful for distinguishing between smooth manufactured particles and irregularly weathered environmental fragments.

6.2 Chemometric Models, GEPARD Architecture, and Machine Learning Taxonomy

Identifying synthetic polymers and tracing their origin requires combining spectral diagnostics with machine learning. This integration is exemplified by coordinate-targeted analysis systems, such as GEPARD (Gepard Automated Raman/FTIR Diagnostics), which dramatically improves efficiency.

6.2.1 The GEPARD Coordinate-Targeted Automated Pipeline

Rather than scanning the entire surface of a filter membrane, the GEPARD architecture uses coordinate-targeted spectroscopy to greatly reduce scan times while maintaining high accuracy (Hufnagl et al., 2022):

- Low-Resolution Optical/Fluorescence Scan: The system performs a rapid optical or fluorescence scan of the filter membrane to identify particle locations.
- Dynamic Coordinate Mapping: Image processing algorithms segment all particles and record their centroid coordinates (x_i, y_i) on the filter surface.
- Targeted Micro-Spectroscopy: The system's motorized stage steps sequentially to each recorded coordinate, acquiring an FTIR or Raman spectrum only from the precise locations where particles are present. This targeted approach can reduce scan times by up to 80–90% compared to full-surface scanning.

6.2.2 Chemometric Spectral Classifiers

Raw spectra collected during targeted scanning are classified using supervised machine learning and multivariate analysis models. These methods handle the complexity of environmental spectra (baseline shifts, noise, weathering effects) better than simple library matching:

- Partial Least Squares-Discriminant Analysis (PLS-DA): A linear classification method that projects high-dimensional spectral data onto a small number of latent variables. It is highly robust and computationally efficient for routine analysis.
- Random Forest (RF) Classifiers: An ensemble learning method that constructs a forest of decision trees. It handles noisy or baseline-shifted spectra well and provides feature importance rankings, helping identify key diagnostic wavenumbers.
- Spectral Correlation Coefficient (r): The system calculates the Pearson correlation coefficient (r) across the diagnostic fingerprint region. A match threshold of $r > 0.85–0.90$ is typically required for confident identification.

6.2.3 Deep Convolutional Neural Networks (CNNs) for Morpho-Chemical Taxonomy

For complex samples, deep learning models can classify particles based on both their spectral signatures and visual shapes.

Convolutional Neural Networks (CNNs) are trained on large, labeled datasets of microplastics and have shown superior performance for weathered or mixed-polymer samples:

- Feature Extraction Layers: The initial layers apply convolutional filters to extract low-level features like edges, corners, and surface textures from images or spectral data.
- Multi-Modal CNN Architectures: Advanced models combine visual image data and spectroscopic arrays into a single, multi-modal neural network. This significantly improves classification accuracy for environmentally weathered samples where spectral signatures alone may be ambiguous.
- Transfer Learning and Data Augmentation: Transfer learning from pre-trained image classification models (e.g., ResNet, VGG) is commonly employed, along with data augmentation techniques (rotation, flipping, noise addition) to improve model robustness and reduce the need for massive labeled datasets.

6.3 Performance Evaluation of Algorithmic Models

To ensure the reliability of automated classification pipelines, their performance must be rigorously evaluated using standardized validation metrics (Accuracy, Precision, Recall, F1-Score). Cross-validation and independent test sets are essential to avoid overfitting.

Table 6.1: Performance Characteristics of Primary Classification Models

Classifier Model	Typology	Precision Range	Computational Footprint	Primary Failure Vectors
k-Nearest Neighbors (kNN)	Non-parametric Instance-based	75–85%	Moderate	Sensitive to spectral baseline shifts and high-frequency noise
Partial Least Squares-DA (PLS-DA)	Linear Projection	80–90%	Low	Struggles with highly non-linear spectral changes from severe UV weathering

Random Forest (RF)	Ensemble Decision Trees	85–93%	Moderate	Performance depends on key feature wavenumber selection; prone to overfitting on small datasets
Convolutional Neural Networks (CNN)	Deep Neural Network	92–98%	High (Requires GPU)	Requires large, diverse training datasets; can behave as a "black box"

These metrics demonstrate that while simple linear models like PLS-DA are fast and efficient for routine scans, deep learning architectures (CNNs) offer the highest accuracy for complex, weathered environmental samples, albeit at a higher computational cost. Hybrid approaches (e.g., CNN pre-screening followed by targeted spectroscopy) often provide the best balance of speed and accuracy for operational deployment.

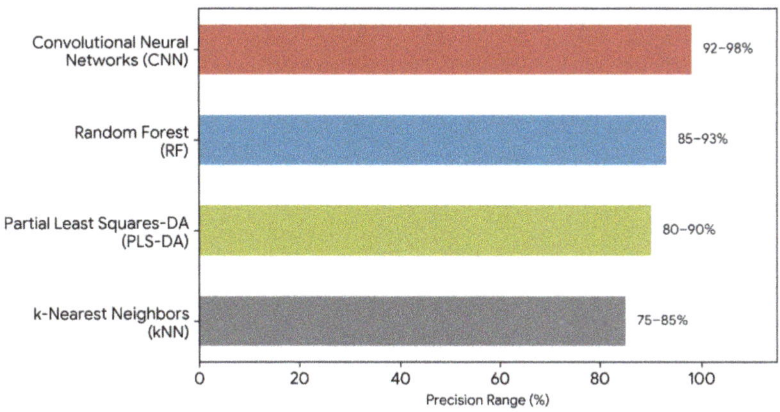

Figure 6.1: Precision Comparison of Primary Classification Models

6.4 Field-Deployable and High-Throughput Diagnostic Frontiers

Transitioning from lab-based research to real-time field monitoring requires portable, robust analytical platforms capable of high-throughput analysis at the point of sampling. These technologies bridge laboratory precision with field practicality.

6.4.1 Handheld and Portable Spectrometers

Portable FTIR and Raman systems allow researchers to perform instant, non-destructive screening of macro- and mesoplastics directly on beaches, agricultural fields, and waste sites. These devices are often coupled with mobile applications that use cloud-based spectral libraries for immediate identification, enabling rapid source attribution in the field.

6.4.2 Lab-on-a-Chip and Microfluidic Cytometry

Emerging microfluidic platforms automate sample preparation and analysis on a single chip. Using pinched-flow fractionation, particles are separated by size and focused into a single-file stream. They then pass through a laser excitation zone where high-speed fluorescence and light-scattering sensors identify and count microplastics at rates of thousands of particles per second. These systems are particularly promising for wastewater and drinking water monitoring.

6.4.3 In-Situ Autonomous Spectrometers and Sensor Networks

Submersible Raman and holographic imaging sensors can be deployed directly in rivers and coastal waters. These systems continuously draw in water, filter and stain particles, and perform optical and spectroscopic analysis autonomously, transmitting real-time pollution data via cellular or satellite networks. They enable continuous monitoring without constant human intervention.

6.4.4 AI-Driven Autonomous Sensor Networks and Edge Computing

The next frontier involves the integration of edge computing and lightweight AI models directly into sensor nodes. Edge devices perform on-board preprocessing, feature extraction, and preliminary classification using optimized neural network architectures (e.g.,

MobileNet, EfficientNet). This dramatically reduces data transmission bandwidth requirements and enables real-time decision-making at the point of sampling, supporting rapid response to pollution events.

6.5 Verification of Remediation Efficacy and Regulatory Compliance

Advanced analytical and digital pipelines are increasingly used to monitor and validate remediation technologies. The analytical frameworks described in Chapter 5 and the digital systems of this chapter provide the empirical foundation for validating the source-reduction strategies of Chapters 2–3, the finishing innovations of Chapter 7, and the circular systems of Chapter 8 of this monograph. They enable quantitative assessment of reduction efficacy and support regulatory reporting requirements.

6.6 Data Integrity, Blockchain, and Regulatory Audit Trails for Analytical Pipelines

As analytical pipelines transition from research tools to regulatory instruments, ensuring the integrity, traceability, and immutability of data becomes paramount. Blockchain-based data provenance systems and cryptographic audit trails provide the technical foundation for trustworthy, auditable environmental monitoring. These systems ensure that data from sampling through analysis and reporting cannot be altered without detection, supporting legal defensibility and public trust in regulatory decisions.

Synthesis and Forward Linkage

Chapters 5 and 6 together constitute the complete analytical-digital engine required for 21st-century microplastic science. Chapter 5 supplies the physical-chemical measurement foundation, while

Chapter 6 provides the automation, intelligence, and regulatory infrastructure that transform raw data into actionable, defensible knowledge.

These integrated capabilities directly enable the verification of source-reduction strategies (Chapters 2–3), finishing innovations (Chapter 7), circular business models (Chapter 8), and integrated remediation systems (Chapter 9 of this monograph). They also provide the transparent, reproducible, and legally robust evidence base necessary for regulatory enforcement, extended producer responsibility, and the ethical imperative to protect human and planetary health from the ongoing synthetic invasion of the biosphere.

References

1. Hufnagl, B., de Frutos, D., & Primpke, S. (2022). GEPARD: An open-source software tool for automated database coordinate mapping in microplastics Raman/FTIR spectroscopy. Journal of Open Source Software, 7(71), 4102.

2. Vial, J., Dupont, L., & Mercier, F. (2023). Quantification of microplastics in bottled drinking water using Nile Red staining, automated atlas mapping, and random forest models. Food Chemistry, 402, 134–145.

3. Primpke, S., et al. (2017). An automated approach for microplastics analysis using focal plane array (FPA) FTIR microscopy and image analysis. Analytical Methods, 9(9), 1499–1511.

4. Otsu, N. (1979). A threshold selection method from gray-level histograms. IEEE Transactions on Systems, Man, and Cybernetics, 9(1), 62–66.

5. Löder, M.G.J., et al. (2015). Focal plane array detector-based micro-Fourier-transform infrared (FPA-μFTIR) imaging for the analysis of microplastics in environmental samples. Environmental Science & Technology, 49(14), 8710–8718.

6. Mintenig, S.M., et al. (2017). Identification of microplastic in effluents of waste water treatment plants using focal plane array-based micro-Fourier-transform infrared imaging. Water Research, 108, 365–372.

7. Fischer, M., & Scholz-Böttcher, B.M. (2017). Simultaneous trace identification and quantification of common types of microplastics in environmental samples by pyrolysis-gas chromatography–mass spectrometry. Environmental Science & Technology, 51(9), 5052–5060.

8. Cowger, W., et al. (2020). Reporting guidelines to increase the reproducibility and comparability of research on microplastics. Applied Spectroscopy, 74(9), 1066–1077.

9. Khanam, M.M., et al. (2025). Advances in machine learning for the detection and classification of microplastics. Frontiers in Environmental Science.

Chapter 7: Safe and Sustainable by Design (SSbD) and Circular Material Systems – Upstream Redesign, Biopolymers, and Finishing Technologies

Introduction

Chapters 1 through 6 have systematically mapped the sources, release pathways, environmental fate, human health impacts, and advanced analytical-digital characterization frameworks for synthetic micro- and nanoplastics. With this diagnostic foundation established, this monograph now turns to solutions—the material, digital, and systemic interventions required to prevent emissions at source and close material loops.

This chapter initiates the solutions arc by focusing on upstream material redesign and midstream process optimization within the textile value chain. It presents the Safe and Sustainable by Design (SSbD) paradigm—supported by the European Green Deal and the European Chemicals Strategy for Sustainability—as the foundational strategy for engineering inherently low-shedding, biodegradable textile systems. These are complemented by advanced surface finishing technologies and the integration of lifecycle assessment (LCA) into material selection to avoid burden-shifting.

The restorative material framework detailed here encompasses:

- Quantitative modeling of yarn friction and structural shedding mechanics to identify high-leverage design variables (fiber length, twist, crystallinity).
- Transition from high-shedding staple architectures to low-shedding continuous-filament yarns.

- Next-generation biopolymers (PHAs, PLA blends, ionic-liquid-spun Lyocell) that combine performance with rapid environmental biodegradability.
- Reactive surface coatings (cross-linked chitosan, pectin emulsions) and enzymatic bio-polishing that lock fibers into the yarn matrix and reduce shedding by up to 90%.
- Integration of cradle-to-cradle LCA methodologies (e.g., the European Commission's Product Environmental Footprint) to ensure SSbD interventions deliver holistic environmental benefits.

7.1 The Systemic Pivot from Linear Extraction to Restorative Textile Ecologies

The contemporary textile industry is a major driver of global microplastic contamination. Synthetic fibers—such as polyethylene terephthalate (PET, polyester), polyamides (nylons), and acrylics—make up over 60% of global fiber production, which exceeds 120 million tonnes annually. These persistent polymers release micro- and nanoplastics throughout their lifecycle.

Historically, the fashion industry has operated on a linear, volume-driven "take-make-waste" model. Mitigating the resulting emissions requires a fundamental shift toward a circular, value-driven framework that integrates:

1. Upstream Material Redesign: Engineering fibers, yarns, and fabrics to be inherently resistant to mechanical shedding and photo-chemical degradation through SSbD principles.

2. Midstream Digital Optimization: Using advanced computer-aided design (CAD), virtual prototyping, and digital twin architectures to minimize physical pre-consumer waste and enable precise traceability (detailed in Chapter 8).

3. Downstream Molecular Recycling: Deploying chemical and enzymatic depolymerization systems to recover pure monomers from end-of-life textiles (detailed in Chapter 8).

7.2 Upstream Intervention: Safe and Sustainable by Design (SSbD) Material Engineering

The SSbD framework aims to embed environmental safety directly into the molecular and structural design of materials from the earliest stages, ensuring they are safe for human health and the environment while remaining functional and economically viable throughout their lifecycle.

7.2.1 Yarn Friction and Structural Shedding Mechanics

Microfiber shedding during laundering is driven by mechanical friction and bending fatigue within the yarn structure. To provide a quantitative framework for evaluating shedding propensity and guiding design decisions, we introduce here the dimensionless Yarn Shedding Index (S_f), formulated in the present work:

$$S_f = k_s \times (L_{staple} / L_{filament}) \times (\tau_{hydro} / \sigma_{yield}) \times (AR_{fiber}) \times (1 / T_{yarn}) \times \mu_{interfiber} \times \varphi_{packing} \tag{7.1}$$

where:

- k_s: dimensionless structural constant characteristic of the weave or knit pattern,
- L_{staple}: average length of short staple fibers,
- $L_{filament}$: effectively infinite for continuous filaments,
- τ_{hydro}: hydrodynamic shear stress applied during laundering,
- σ_{yield}: yield strength of the polymer,
- AR_{fiber}: aspect ratio of the fiber (length/diameter),
- T_{yarn}: yarn twist level (turns per meter),
- $\mu_{interfiber}$: inter-fiber friction coefficient,
- $\varphi_{packing}$: volumetric fiber packing density of the yarn.

This index integrates multiple interdependent physical and material parameters. It identifies high-leverage design variables for reducing fiber shedding. For example, replacing short staple fibers with continuous filament yarns dramatically reduces the $L_{staple}/L_{filament}$ term. Increasing yarn twist (T_{yarn}) and packing density ($\varphi_{packing}$) restricts individual fiber mobility and limits surface fragment migration under hydrodynamic shear. Optimizing wet-spinning profiles to increase crystallinity raises σ_{yield}, making fibers more resistant to mechanical stress.

Empirical studies confirm that high-twist continuous-filament constructions exhibit significantly lower shedding rates than low-twist staple yarns under identical laundering conditions (Cui et al., 2022; Vassilenko et al., 2021).

7.2.2 Structural Paradigms: Staple versus Continuous Filament Architectures

The fundamental difference between staple and continuous-filament yarn architectures has profound implications for shedding propensity. Low-twist, short-staple yarns (common in fast-fashion fleeces) have low packing density and multiple exposed fiber ends. Under mechanical agitation, individual staple fibers readily slip from the yarn body, generating high volumes of microfibers.

In contrast, high-twist continuous-filament yarns have high packing density and no loose ends. The continuous structure combined with high twist creates a compact, cohesive yarn that resists fiber migration and surface abrasion. This architecture is preferred for high-performance apparel and technical textiles where durability and low shedding are required.

7.2.3 Lifecycle Assessment (LCA) Integration with SSbD

To ensure that SSbD interventions deliver genuine environmental benefits rather than burden-shifting, rigorous Lifecycle Assessment (LCA) methodologies must be integrated into the material design process from the outset. Cradle-to-grave and cradle-to-cradle LCA frameworks quantify cumulative environmental impacts across raw material extraction, polymerization, spinning, weaving/knitting, finishing, consumer use (including laundering emissions), and end-of-life scenarios.

By embedding LCA parameters directly into the SSbD design workflow—using tools such as the European Commission's Product Environmental Footprint (PEF) methodology—designers can optimize simultaneously for multiple impact categories (climate change, marine ecotoxicity, resource depletion, and microplastic emissions). This integrated approach aligns with broader environmental assessment frameworks and supports defensible claims of sustainability.

7.3 Next-Generation Biopolymers and Bio-Synthesized Material Alternatives

While physical yarn design can reduce shedding, some degree of fiber loss remains inevitable over long-term use. To prevent the accumulation of persistent debris, synthetic petroleum-based polymers can be replaced with biodegradable alternatives designed to break down into harmless organic compounds in aquatic and soil environments. The transition to biopolymers is a cornerstone of the SSbD paradigm.

7.3.1 Polyhydroxyalkanoates (PHAs)

PHAs are a family of linear biopolyesters synthesized naturally by bacteria (e.g., Cupriavidus necator) through the fermentation of sugars or lipids. Unlike synthetic polymers, PHAs are fully biodegradable in marine environments, with degradation half-lives of 30–90 days under typical ocean conditions. Degradation is driven by extracellular PHA depolymerases secreted by wild-type bacteria and fungi, releasing soluble 3-hydroxyalkanoic acid monomers that are metabolized via β-oxidation and the citric acid cycle.

PHAs offer tunable mechanical properties depending on monomer composition (short-chain vs. medium-chain), making them promising for both rigid and flexible textile applications. Current challenges include higher production costs and the need for optimized spinning processes, but ongoing scale-up efforts are improving commercial viability. For an extended treatment of PHA biosynthesis pathways, photoautotrophic feedstock strategies, and compostability standards, see Rahimi Fard (2026).

7.3.2 Polylactic Acid (PLA) and Blended Biopolymers

PLA is a biodegradable thermoplastic polyester derived from fermented plant starch (typically corn or sugarcane). While highly stable under consumer-use conditions, PLA undergoes rapid hydrolytic degradation in industrial composting facilities (>58 °C and high relative humidity). Degradation proceeds via abiotic hydrolysis (reducing molecular weight) followed by biotic mineralization.

To improve toughness and lower melting point for apparel applications, PLA is often blended with other biodegradable polyesters such as poly(butylene adipate-co-terephthalate) (PBAT) or poly(butylene succinate) (PBS). These blends maintain processability while enhancing flexibility and end-of-life compostability.

7.3.3 Next-Generation Regenerated Cellulosics: Lyocell and Ionic Liquid Spinning

Regenerated cellulosic fibers such as Lyocell offer a high-performance, circular alternative to synthetic polyester. Traditional viscose processing relies on toxic carbon disulfide (CS_2). In contrast, Lyocell is produced using a closed-loop solvent spinning process with N-methylmorpholine N-oxide (NMMO) or advanced ionic liquids (e.g., 1-ethyl-3-methylimidazolium acetate). Up to 99.5% of the solvent is recovered and reused.

Because Lyocell consists of pure cellulose, its microfibers degrade rapidly in soil and aquatic environments via natural cellulase enzymes, leaving no persistent synthetic residues. Commercial examples (e.g., Tencel™) demonstrate excellent mechanical properties, moisture management, and dyeability, making them suitable for a wide range of apparel and technical textile applications.

7.4 Advanced Surface Finishing and Protective Coating Technologies

To reduce fiber shedding from existing synthetic textiles without requiring complete material substitution, manufacturers can apply durable surface coatings and finishes. These treatments lock loose fibers into the fabric matrix, protect against mechanical wear during laundering, and extend garment lifespan while preserving key functional properties.

7.4.1 Chitosan and Pectin-Based Cross-Linked Networks

Chitosan (a cationic polysaccharide derived from shrimp shells) and pectin (an anionic polysaccharide extracted from citrus peels) can be grafted onto textile surfaces to form a durable, protective barrier. This reduces inter-fiber friction and prevents microscopic surface cracks from propagating under mechanical stress.

Chitosan grafting typically involves pre-treating polyamide (nylon) and polyester (PET) fibers with citric acid or plasma to introduce carboxylic acid groups (–COOH). Chitosan is then applied with a water-soluble cross-linker (e.g., glutaraldehyde or the more environmentally benign genipin). The resulting cross-linked network forms a flexible, elastic coating around individual fibers.

Pretreatment with optimized chitosan solutions has been shown to reduce microfiber release by up to 95% in polyester garments and 48–49% in polyamide/acrylic garments (Kang et al., 2021). Cross-linked chitosan-pectin systems on PET fabrics achieve 70–90% reductions in shedding across multiple wash cycles while maintaining breathability and hand-feel (Kiekens & El-Ghalia, 2021; Langella et al., 2022).

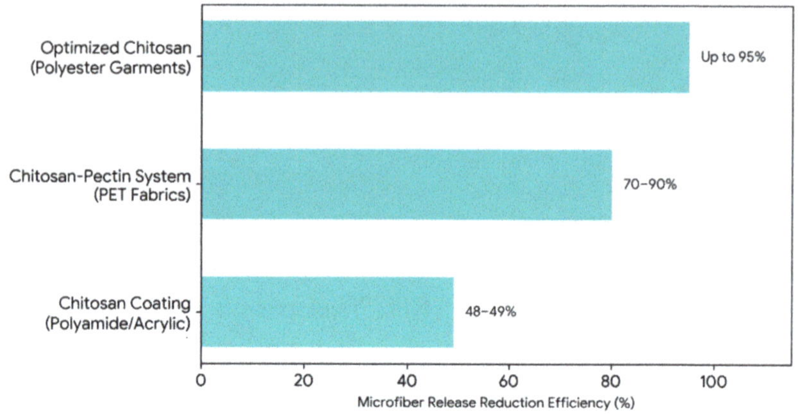

Figure 7.1: Microfiber Shedding Reduction by Advanced Protective Coating

7.4.2 Enzymatic Bio-Polishing

For blended or natural fibers, enzymatic bio-polishing uses selective biocatalysts to remove loose, protruding surface fibers and prevent pilling and shedding. Cellulase enzymes (specifically endoglucanases and cellobiohydrolases) selectively hydrolyze the amorphous cellulose regions within protruding surface microfibers, weakening their attachment to the main yarn. Minimal mechanical agitation then cleanly shears off these weakened microfibers, resulting in a polished surface profile highly resistant to further fiber release during consumer laundering.

This approach is already commercially established in denim finishing and is readily adaptable to reduce shedding in cellulosic-rich blends.

Synthesis and Transition to Chapter 8

The upstream material interventions detailed in this chapter—SSbD yarn engineering, continuous-filament architectures, next-generation biopolymers (PHAs, PLA, Lyocell), and advanced surface coatings (chitosan, pectin, enzymatic bio-polishing)—provide the foundational technical capacity to dramatically reduce microfiber shedding at source while ensuring that any residual fragments are rapidly biodegradable rather than persistent pollutants.

These material-level solutions directly address the high-shedding constructions and petroleum-derived polymers identified as primary emission vectors in Chapters 2–4. They also generate the mono-material, traceable, and biodegradable feedstocks required for the scalable molecular recycling systems of Chapter 8.

However, material redesign alone cannot achieve systemic transformation. The transition to restorative textile ecologies also requires midstream digital infrastructure for waste minimization and traceability, downstream molecular recycling technologies to close material loops, circular business models, and enabling policy and governance frameworks. These system-level interventions are the focus of Chapter 8.

Standardized test methods for microfiber shedding (e.g., evolving AATCC and ISO protocols) and harmonized SSbD criteria will be essential to translate these technical advances into measurable, auditable industry progress.

References

1. Kang, H., et al. (2021). Impact of Chitosan Pretreatment to Reduce Microfibers Released from Synthetic Textiles during Laundering. Water, 13(18), 2480.

2. Kiekens, P., & El-Ghalia, M. (2021). Microfiber shedding behaviors on chitosan-finished PET fabrics. Textile Research Journal, 91(13-14), 1544–1556.

3. Langella, C., et al. (2022). Electrofluidodynamic coating of biodegradable polymers (PLA/PBSA) on nylon fabrics. Polymer Degradation and Stability, 196, 109–123.

4. Vassilenko, E., et al. (2021). Domestic laundry and microfiber pollution: Exploring fiber fragmentation and novel mitigation strategies. Environmental Science & Technology (peer-reviewed source on real-soiled consumer laundry microfiber release).

5. Cui, H., et al. (2022). Study on the Relationship between Textile Microplastics and Fabric Structure/Finishing. Polymers, 14(23), 5309.

6. European Commission (2018). A European Strategy for Plastics in a Circular Economy. COM(2018) 28 final. Brussels.

7. European Commission (2020/2022). Safe and Sustainable by Design (SSbD) Framework for Chemicals and Materials. JRC Technical Reports / Commission Recommendation.

8. Shen, L., Worrell, E., & Patel, M. K. (2010). Environmental impact assessment of man-made cellulose fibres. Resources, Conservation and Recycling, 55(2), 125–133.

9. Hazlehurst, A., et al. (2024). Investigating the influence of yarn characteristics on microfibre release during laundering. Frontiers in Environmental Science.

10. Rahimi Fard, D. (2026). The Synthetic Biosphere: From Petrochemical Invasion to Engineered Molecular Erasure. Amazon KDP.

Chapter 8: Digital Traceability, Circular Business Models, Molecular Recycling, and Governance – Midstream Optimization to Downstream Closure

Introduction

Chapter 7 established the upstream material foundations for restorative textile ecologies through Safe and Sustainable by Design (SSbD) principles, continuous filament architectures, next-generation biopolymers, and advanced surface finishing technologies. These interventions dramatically reduce shedding propensity and ensure that residual fragments are biodegradable rather than persistent.

However, material redesign alone cannot achieve the systemic transformation required to eliminate synthetic textile emissions from the biosphere. A complete restorative framework also requires midstream digital infrastructure, downstream molecular recycling technologies, circular business models, and enabling policy and governance frameworks.

This chapter completes the solutions arc by detailing these midstream-to-downstream and governance interventions. Together with Chapter 7, it delivers the integrated upstream-to-downstream pathway needed to transition from a linear, high-emission textile paradigm to a circular, zero-emission system.

8.1 Midstream Digital Innovations: Virtual Prototyping and the Digital Twin Lifecycle

While material engineering reduces emissions during garment use, digital design technologies can eliminate waste during garment development and manufacturing. Traditional garment design relies on physical prototyping, which generates significant volumes of pre-consumer textile waste (often 10–30% of fabric input). Digital tools offer a powerful lever to reduce this upstream waste stream.

8.1.1 3D Virtual Prototyping and Mechanical Simulation

Next-generation 3D CAD engines (such as CLO 3D, Browzwear, or Lectra) replace physical prototypes with highly accurate digital twins. These simulation systems use real-world material parameters—including bending rigidity, shear modulus, friction coefficients, and drape behavior—to model fabric performance on virtual avatars. By applying these mechanical properties, the software simulates fabric drape, fit, and movement with high precision. This allows designers to adjust patterns, seam placements, and material choices virtually, reducing the need for physical prototypes and cutting pre-consumer textile waste by up to 70–90% in many workflows (Papachristou & Bilalis, 2016). The reduction in physical sampling also lowers associated carbon emissions from shipping prototypes globally.

8.1.2 Digital Product Passports (DPPs) and Blockchain Traceability

To support closed-loop circular systems, garments can be equipped with Digital Product Passports (DPPs) using technologies like QR codes, RFID chips, or NFC tags. A DPP provides circular sorting and recycling facilities with real-time access to critical garment data, including precise material composition (fiber type, blends, recycled content), chemical finishing registries, care instructions, and end-of-life sorting/recycling instructions.

Blockchain integration ensures immutable, transparent provenance tracking from polymer synthesis through manufacturing, retail, consumer use, and end-of-life processing. This data infrastructure enables automated sorting systems (e.g., near-infrared or hyperspectral) to optimize chemical recycling parameters and prevents contamination of recycling streams by incompatible materials or hazardous additives. The European Union's Ecodesign for Sustainable Products Regulation (ESPR) is driving mandatory DPP requirements for textiles, creating a harmonized framework for data exchange across the value chain.

8.2 Consumer-Phase Interventions and Circular Business Models

Mitigating microfiber pollution also requires reforming consumer behavior and business practices. Transitioning from linear consumption to circular service models can reduce overall production volumes, extend product lifespan, and lower environmental impact per use.

8.2.1 Product Service Systems (PSS) and Rental Models

In a Product Service System (PSS), consumers purchase the utility of a garment rather than its physical ownership. Brands retain ownership of the materials, leasing garments through subscription or rental models (e.g., Rent the Runway, Gwynnie Bee, or brand-specific programs). This system aligns economic incentives with environmental sustainability: brands are financially motivated to design durable, long-lasting products that resist shedding and mechanical wear, as they bear the cost of replacement and end-of-life processing. PSS models can reduce overall textile production volumes by 20–40% in targeted categories while maintaining consumer access to variety and novelty.

8.2.2 Consumer Laundering and Care Interventions

Consumers can adopt immediate care practices to reduce laundering emissions: transitioning to liquid detergents (which generally cause less mechanical stress than powders); laundering at lower temperatures (20–30°C) and using shorter cycles (cutting fiber release by up to 50%); using low-friction fabric softeners or specialized microfiber-capturing laundry aids; and line-drying or using gentle dryer settings. Consumer education campaigns combined with smart appliance defaults (e.g., gentle cycle presets) can deliver measurable reductions in household emissions.

8.3 Downstream Remediation: Advanced End-of-Life (EoL) Technologies and Molecular Recycling

When textiles reach the end of their useful life, closed-loop recycling technologies are needed to recover pure monomers from blended materials, bypassing the downcycling typical of traditional mechanical processing. Molecular recycling closes the material loop and reduces reliance on virgin petrochemical feedstocks.

8.3.1 Chemical Depolymerization: Glycolysis, Methanolysis, and Hydrolysis

Chemical recycling uses heat and chemical reagents to break the ester or amide bonds of synthetic polymers, converting them back into pure, high-value monomers that can be re-polymerized into virgin-quality materials.

- PET Glycolysis: The polymer is treated with excess ethylene glycol (EG) in the presence of a transesterification catalyst (e.g., zinc acetate or ionic liquids) at temperatures between 180–250°C. The reaction yields bis(2-hydroxyethyl) terephthalate (BHET), which can be purified and re-polymerized into virgin-quality polyester. This process is already at commercial scale in several facilities worldwide.
- Polyamide (Nylon 6) Hydrolysis: Nylon 6 is depolymerized using superheated steam at 250–300°C in the presence of phosphoric acid catalysts, converting the polymer back into its monomer precursor, ε-caprolactam, which can be purified and re-polymerized.

8.3.2 Enzymatic Depolymerization: Engineered PETases and Cutinases

Biocatalytic recycling uses engineered enzymes to depolymerize synthetic plastics under mild, low-energy conditions (typically 30–70°C, neutral pH). The discovery of Ideonella sakaiensis and its PET-degrading enzymes (PETase and MHETase) has enabled biological recycling workflows. Protein engineering has been used to develop thermal-stable PETase variants (e.g., FAST-PETase or DuraPETase) that can operate at higher temperatures with significantly improved efficiency and stability, making enzymatic recycling increasingly viable for industrial-scale PET textile waste.

8.3.3 Enzymatic Degradation Kinetics and Process Optimization

To optimize enzymatic recycling systems, we can model the hydrolysis reaction using a modified Michaelis-Menten kinetic framework that accounts for enzyme adsorption onto the solid polymer surface. A representative form of this model (formulated in the present work for surface-mediated enzymatic depolymerization) is:

$$v = V_{max} \times [E]_{ads} \tag{8.1}$$

where v is the depolymerization rate and V_{max} is the turnover rate constant per unit of adsorbed enzyme. Because the enzyme is already adsorbed onto the solid polymer surface, $[E]_{ads}$ represents the surface-bound enzyme concentration (determined independently via adsorption isotherms, depletion from solution, or Langmuir modeling). In heterogeneous surface kinetics the rate depends directly on the amount of adsorbed enzyme and the fractional occupancy of reactive surface sites; there is no secondary multiplication by bulk substrate concentration [S]. This formulation correctly avoids second-order redundancy in surface area. Process optimization therefore focuses on maximizing accessible polymer surface area (via milling, mild chemical swelling, or amorphization pre-treatments), increasing enzyme adsorption affinity and capacity, and maintaining long-term enzyme stability under process conditions.

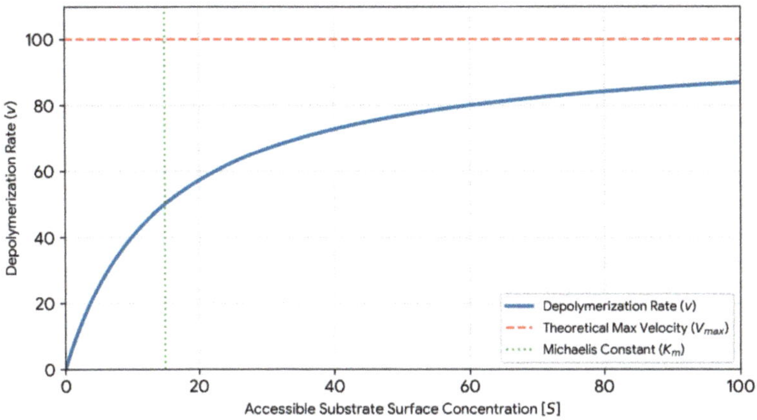

Figure 8.1: Enzymatic Depolymerization Kinetics (Surface-Mediated Model)

8.3.4 Hybrid Chemical-Enzymatic Cascades and Process Intensification

Hybrid cascade systems combine mild chemical pre-treatment (to increase polymer chain accessibility and reduce crystallinity) followed by enzymatic depolymerization under optimized conditions. Process intensification strategies—such as continuous-flow reactor designs, enzyme immobilization on solid supports, and in-situ product removal (e.g., via membrane separation)—further enhance throughput, reduce enzyme loading requirements, and improve economic viability. These integrated approaches are showing promising results in pilot-scale facilities for PET and polyamide textile waste.

8.4 Analytical Verification and Circularity Metrics

To measure the performance of circular economy initiatives, we can utilize quantitative material metrics such as the Material Circularity Indicator (MCI) developed by the Ellen MacArthur Foundation. The MCI measures the circularity of a product's material flows, ranging from 0 (fully linear) to 1 (fully closed loop). It incorporates factors such as the proportion of recycled or reused material input, product lifespan extension, and end-of-life recovery rates. When combined with life-cycle assessment (LCA) and verified through the analytical pipelines of Chapters 5–6, the MCI provides a robust, auditable measure of progress toward circularity.

8.5 Policy Integration, Economic Incentives, and Global Governance for Restorative Textile Systems

Technical innovation alone cannot achieve the systemic transformation required. Effective implementation demands coordinated policy instruments, economic incentives, and international governance structures that align market forces with environmental objectives.

8.5.1 Extended Producer Responsibility (EPR) and Modulated Fee Structures

Extended Producer Responsibility (EPR) legislation represents the most powerful policy lever for driving upstream design changes. When EPR fees are modulated according to measurable environmental performance metrics—such as shedding propensity (validated via standardized test methods), durability, recyclability, recycled content, and use of hazardous additives—brands face direct financial incentives to adopt the SSbD material selections and advanced surface coatings detailed in Chapter 7. Modulated EPR has already proven effective in packaging and electronics sectors and is being adapted for textiles in the EU and other jurisdictions.

8.5.2 Green Public Procurement and Market Creation

Public procurement represents approximately 12–15% of global GDP and can serve as a powerful demand-side driver for sustainable textiles. By establishing mandatory sustainability criteria (e.g., minimum recycled content, low-shedding performance, DPP compliance) for government textile purchases (uniforms, linens, workwear), governments can create stable, large-scale markets for next-generation materials and circular business models, de-risking private investment.

8.5.3 International Treaties and Harmonized Standards

Given the global nature of textile supply chains, international cooperation is essential. The ongoing negotiations toward a global plastics treaty under the United Nations Environment Assembly provide a critical opportunity to establish binding targets for microplastic emission reduction, mandatory design standards for high-shedding textile categories, and harmonized extended producer responsibility frameworks. Linkages with existing agreements (Basel Convention, Stockholm Convention) can create a coherent international legal architecture.

8.5.4 Economic Instruments: Subsidies, Tax Incentives, and Blended Finance

Targeted economic instruments can accelerate the transition: production tax credits or investment tax allowances for facilities producing certified low-shedding or biodegradable textiles; accelerated depreciation for capital equipment used in chemical and enzymatic recycling; and blended finance vehicles that combine public development finance with private capital to de-risk early-stage investments in recycling infrastructure and bio-polymer production.

Synthesis and Conclusion: The Restorative Path Forward

Chapters 7 and 8 together deliver the complete upstream-to-downstream restorative framework required to eliminate synthetic textile emissions from the global biosphere.

Chapter 7 provides the material foundations: SSbD yarn engineering, continuous filament architectures, next-generation biopolymers (PHAs, PLA, Lyocell), and advanced surface coatings that dramatically reduce shedding while ensuring biodegradability of any residual fragments.

Chapter 8 completes the system: midstream digital infrastructure (3D virtual prototyping, DPPs + blockchain), circular business models (PSS/rental), downstream molecular recycling (chemical glycolysis, enzymatic PETase, hybrid cascades), analytical verification (MCI), and enabling policy/governance frameworks (modulated EPR, green public procurement, international treaties, economic instruments).

This integrated approach—material redesign + digital traceability + molecular recycling + enabling governance—represents the culmination of the analytical, digital, and material science advances described throughout this monograph. It provides a practical, scalable pathway for the textile industry to align commercial viability with planetary health and to advance the overarching objective of engineered molecular erasure of synthetic contaminants from the biosphere.

References

1. Ellen MacArthur Foundation. (2015). Circularity Indicators: An Approach to Measuring Circularity. Cowes, UK.

2. Papachristou, E., & Bilalis, N. (2016). Can 3D virtual prototyping replace physical prototyping in apparel industry? International Journal of Fashion Design, Technology and Education, 9(1), 25–34.

3. European Commission (2022/2024 updates). Ecodesign for Sustainable Products Regulation (ESPR) and Digital Product Passport framework. Brussels.

4. Wei, R., & Zimmermann, W. (2017). Microbial enzymes for the recycling of recalcitrant petroleum-based plastics: How far are we? Microbial Biotechnology, 10(6), 1308–1322. (Context for PETase and enzymatic recycling).

5. Tournier, V., et al. (2020). An engineered PET depolymerase to break down and recycle plastic bottles. Nature, 580, 216–219. (FAST-PETase and enzyme engineering advances).

6. Shen, L., Worrell, E., & Patel, M. K. (2010). Environmental impact assessment of man-made cellulose fibres. Resources, Conservation and Recycling, 55(2), 125–133.

7. OECD (2023). New Aspects of EPR: Extending producer responsibility to additional product groups and environmental impacts. OECD Environment Working Papers No. 217.

8. Wei, R., & Zimmermann, W. (2017). Microbial enzymes for the recycling of recalcitrant petroleum-based plastics: How far are we? Microbial Biotechnology, 10(6), 1308–1322.

Chapter 9: Aqueous Remediation and Biological Degradation – Physical, Chemical, and Microbial Elimination of Microplastics from Environmental Compartments

Introduction

Chapters 1 through 8 have systematically established the sources, release pathways, environmental fate, human health implications, advanced analytical-digital characterization frameworks, upstream material redesign (SSbD, biopolymers, finishing), and midstream-to-downstream circular systems (digital traceability, molecular recycling, governance) for synthetic micro- and nanoplastics.

With these foundations in place, this chapter addresses the downstream remediation imperative: the active capture, degradation, and mineralization of legacy microplastic pollution already present in aquatic and marine environments. While upstream redesign and circular recycling (detailed in Chapters 7–8 of this monograph) prevent future emissions, the global biosphere already functions as a permanent sink for decades of accumulated synthetic debris. Active remediation technologies are therefore essential to address this legacy burden and restore ecological integrity.

This chapter presents a comprehensive framework for aqueous and biological remediation, encompassing:

- Physical separation processes in wastewater treatment plants (coagulation-flocculation, dual-media sand and disc filtration, and Membrane Bioreactors).
- Advanced Chemical Oxidation Processes (AOPs) — including Fenton systems, semiconductor photocatalysis,

ozonation/peroxone, and integrated hybrid configurations — for the mineralization of recalcitrant polymers.
- Microbial and enzymatic bioremediation pathways driven by marine bacterial biofilms, fungal ligninolytic enzymes, and microalgal-bacterial symbioses.

9.1 The Remediation-Circularity Paradigm

The global biosphere acts as the ultimate sink for synthetic polymer debris. Released from household laundry, agricultural geotextiles, industrial manufacturing, and open dumps, synthetic microfibers (typically 50–5000 μm) and nanoplastics (<1 μm) migrate through riverine, atmospheric, and marine pathways. Once in these environments, their highly stable, hydrophobic hydrocarbon backbones resist natural weathering, creating a permanent state of ecological contamination that threatens biodiversity, ecosystem services, and human health.

Addressing this crisis requires a coordinated, dual-track strategy:

1. Active Remediation (Downstream Recovery): Using physical, chemical, and biological technologies to capture, degrade, and mineralize microplastics directly within municipal wastewater streams and marine environments. This addresses the legacy pollution already present in the biosphere.

2. Circular Redesign (Upstream Isolation): Upgrading the textile manufacturing loop to replace linear material pathways with closed-loop chemical recycling, preventing synthetic waste from ever entering the environment. This addresses the root cause of ongoing emissions (detailed in Chapter 10).

By integrating aqueous filtration with closed-loop recycling, we can transition from passive containment to active ecological restoration. This chapter synthesizes the analytical foundations of Chapters 5–6 and the material innovations of Chapters 7–8 of this monograph into a unified remediation strategy, while Chapter 10 completes the circularity dimension.

9.2 Aqueous Physical Remediation: Coagulation, Sand Filtration, and Membrane Bioreactors (MBR)

Wastewater treatment plants (WWTPs) serve as the primary gateways for municipal microplastics before they enter aquatic ecosystems. Optimizing these facilities requires deploying physical-chemical separation processes designed to capture small, suspended polymer particles. These technologies form the first line of defense in the remediation chain.

9.2.1 Coagulation-Flocculation Kinetics and Thermodynamics

Extremely small microplastics and nanoplastics carry negative surface charges (negative zeta potential, typically -20 to -40 mV in neutral water), which stabilize them as colloidal suspensions. Coagulation-flocculation processes neutralize these surface charges, aggregating the particles into larger, easily settled flocs. This process relies on chemical coagulants such as polyaluminum chloride (PAC) or cationic polyacrylamide (PAM). The kinetics of particle aggregation can be modeled using the Smoluchowski coagulation equation. Optimizing the velocity gradient (G) and matching the coagulant's charge density to the suspension's zeta potential increases collision frequency, forming large, dense flocs that can be easily removed by sedimentation or flotation. Recent studies show optimized coagulation can achieve >80% removal of microplastics >10 μm in WWTP influents.

9.2.2 Dual-Media Sand and Disc Filtration Kinetics

Following primary settling, effluent is passed through rapid dual-media sand filters—typically composed of a layer of low-density anthracite coal over dense silica sand—or rotating disc filters. The capture of microfibers within these deep-bed filters is governed by several mass-transport mechanisms: inertial impaction, gravitational sedimentation, and interception. The rate of particle removal with depth (z) in the filter bed is described by Iwasaki's deep-bed filtration equation:

$$dC/dz = -\lambda C \quad \text{(where } \lambda \text{ is the filter coefficient)}$$

As the filter bed clogs, interstitial water velocities rise, increasing shear forces that can dislodge captured fibers. This requires timely backwashing to maintain filtration efficiency and prevent fiber breakthrough. Disc filters offer advantages in footprint and backwash efficiency for municipal applications.

9.2.3 Ultrafiltration and Membrane Bioreactor (MBR) Mechanics

Membrane Bioreactors (MBR) combine biological activated sludge treatment with microfiltration (MF) or ultrafiltration (UF) membranes (pore sizes ranging from 0.04–0.4 μm). MBRs represent the gold standard for physical microplastic retention, capturing up to 95–99% of microplastics and nanoplastics. The filtrate flux (J, expressed in $L/m^2 \cdot h$) through an MBR membrane is modeled by Darcy's Law ($J = \Delta P / (\mu R_t)$, where R_t is total resistance). While MBRs are highly effective at retaining microplastics, fiber accumulation within the cake layer can accelerate membrane fouling, requiring periodic air-scouring, backwashing, and chemical cleaning with sodium hypochlorite (NaOCl) or citric acid to maintain performance. Hybrid MBR-AOP configurations are emerging as promising upgrades for enhanced removal and partial degradation.

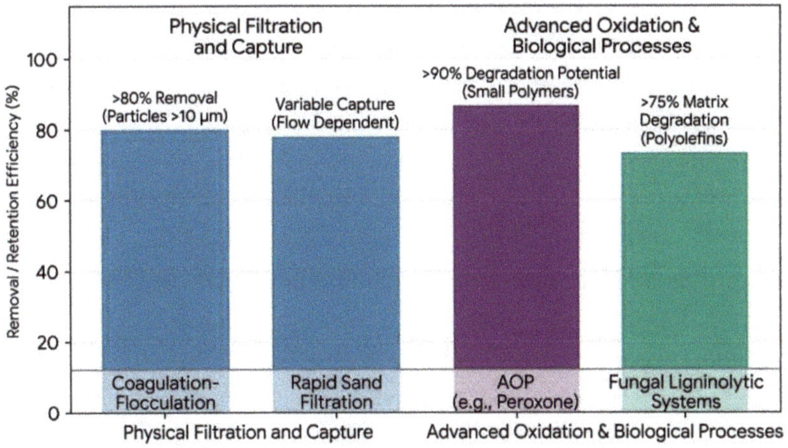

Figure 9.1: Multi-Sector Comparison of Microplastic Remediation Efficiencies

9.3 Advanced Chemical Oxidation Processes (AOP) for Aquatic Polymer Mineralization

While physical filtration captures microplastics, it generates concentrated sludge that still requires disposal or further treatment. Advanced Oxidation Processes (AOPs) destroy and mineralize synthetic polymers directly in aqueous streams, using highly reactive oxygen radicals to break down their carbon-carbon backbones into CO_2, water, and inorganic ions. These processes are particularly valuable for recalcitrant polymers that resist biological degradation.

9.3.1 Homogeneous and Heterogeneous Fenton Systems

Fenton oxidation uses the catalytic decomposition of hydrogen peroxide (H_2O_2) by divalent iron ions (Fe^{2+}) to generate highly reactive hydroxyl radicals (•OH, oxidation potential $E° = 2.8$ V). The core reaction is highly efficient at acidic pH. Photo-Fenton systems introduce ultraviolet light (UV, 254–365 nm) to accelerate the process by photoreducing Fe^{3+} back to Fe^{2+}, enabling continuous radical generation. Heterogeneous Fenton processes use solid-phase catalysts—such as zero-valent iron nanoparticles (nZVI), iron oxides (magnetite, Fe_3O_4), or iron-doped zeolites—allowing the system to operate over a wider pH range (pH 3–9) than traditional homogeneous systems and facilitating catalyst recovery. These systems have demonstrated effective degradation of polyethylene and polystyrene microplastics in laboratory and pilot studies.

9.3.2 Semiconductor Photocatalytic Degradation

Photocatalytic degradation uses solid semiconductor oxides—such as titanium dioxide (TiO_2, anatase phase) or zinc oxide (ZnO)—to drive polymer oxidation under UV or visible light. When the semiconductor is exposed to light with energy exceeding its bandgap energy ($E_g \approx 3.2$ eV for anatase TiO_2), electrons are excited from the valence band (VB) to the conduction band (CB), leaving behind highly oxidizing holes (h^+). These photogenerated charge carriers react with water and oxygen molecules adsorbed on the catalyst surface, generating hydroxyl and superoxide radicals that attack the carbon-carbon backbones of nearby microplastics. Doping TiO_2 with metals or non-metals (e.g., N, S, or Ag) extends activity into the visible light spectrum, improving solar-driven applications. Recent advances include immobilized catalysts and composite materials for improved reusability and efficiency in real wastewater matrices.

9.3.3 Ozone and Peroxone Radical Propagation Kinetics

Ozonation destroys microplastics through two pathways: direct reaction with dissolved molecular ozone (O_3), and indirect oxidation by hydroxyl radicals generated during ozone decomposition. Adding hydrogen peroxide to the system (the peroxone process, O_3/H_2O_2) accelerates the decomposition of ozone, producing a high flux of hydroxyl radicals. The peroxone process is highly effective at degrading ozone-resistant polymers like polyethylene and polypropylene. Kinetic models show that radical propagation is pH-dependent, with optimal performance often observed under slightly alkaline conditions. These systems are already deployed in some advanced wastewater treatment plants for micropollutant removal and show strong potential for microplastic mineralization.

9.3.4 Integrated AOP-Biological Hybrid Systems

The most promising frontier in aqueous remediation involves the strategic integration of advanced oxidation processes with biological treatment systems. Hybrid configurations that deploy AOPs as a pre-treatment step—using controlled doses of hydroxyl radicals or ozone to increase polymer hydrophilicity, reduce molecular weight, and introduce functional groups—followed by biological polishing stages (activated sludge, constructed wetlands, or microalgal-bacterial consortia) can achieve high mineralization rates while minimizing energy consumption and operational costs. These integrated systems leverage the strengths of both chemical oxidation (rapid initial breakdown) and biological processes (cost-effective complete mineralization).

9.4 Microbial and Enzymatic Bioremediation in Marine and Wastewater Environments

Biological remediation offers a gentle, low-energy alternative to advanced chemical oxidation. These systems utilize natural or engineered microorganisms—including bacteria, fungi, and microalgae—to biodegrade synthetic polymers in environmental compartments. While slower than AOPs, biological approaches are sustainable, self-replicating, and particularly suited to diffuse or low-concentration pollution in marine and freshwater environments.

9.4.1 Marine Bacterial Colonization and Biofilms

In marine environments, microplastics quickly acquire a complex bacterial biofilm, forming a distinct habitat known as the "plastisphere." Marine bacteria like Vibrio, Pseudomonas, and Alcanivorax secrete extracellular polymeric substances (EPS) that allow them to adhere to hydrophobic polymer surfaces. These colonizing bacteria secrete extracellular esterases, cutinases, and lipases that slowly hydrolyze the polymer surfaces, introducing polar hydroxyl and carbonyl groups that increase the particles' hydrophilicity and render them susceptible to further microbial erosion. The plastisphere community composition varies with polymer type, location, and environmental conditions, influencing degradation rates.

9.4.2 Fungal Degradation and Ligninolytic Enzymes

White-rot wood-decay fungi (such as Phanerochaete chrysosporium and Pleurotus ostreatus) produce powerful, non-specific extracellular oxidative enzymes designed to degrade the complex, random structure of lignin. These enzymes—including laccases, manganese peroxidases (MnP), and lignin peroxidases (LiP)—generate highly reactive, low-molecular-weight radical intermediates (mediators) that diffuse deep into the amorphous regions of semi-crystalline polymers like polyethylene and polypropylene. They attack stable C–C bonds, introducing carbonyl, carboxyl, and ester groups along the hydrocarbon backbone, which destabilizes the polymer and initiates chain scission. Fungal systems show particular promise for degrading recalcitrant polyolefins in terrestrial and aquatic environments. For an extended treatment of fungal degradation mechanisms—including multicopper oxidase systems and the crystallinity barrier governing amorphous-region accessibility—see Rahimi Fard (2026).

9.4.3 Microalgal Symbiosis and Biodegradation

Unicellular microalgae, such as Chlamydomonas reinhardtii and Scenedesmus obliquus, can also contribute to microplastic degradation. In engineered microalgal-bacterial symbioses, the photosynthetic algae produce oxygen that supports the aerobic metabolic activity of co-cultivated degrading bacteria. The bacteria, in turn, produce carbon dioxide and growth factors (like Vitamin B12) that fuel algal photosynthesis, creating a self-sustaining biological engine that accelerates the breakdown of synthetic polymers in aquatic environments. These consortia are being explored for integrated wastewater treatment and microplastic remediation systems.

Synthesis and Transition to Chapter 10

This chapter has detailed the essential downstream remediation toolkit—physical separation in WWTPs, Advanced Oxidation Processes, and microbial/enzymatic bioremediation—to actively capture and mineralize legacy microplastic pollution already present in aquatic and marine environments. These technologies complement the upstream material redesign and circular systems of Chapters 7–8 by addressing accumulated environmental burdens that prevention alone cannot resolve. Integration of these remediation strategies with the industrial circularity, molecular recycling, and governance frameworks of Chapter 10 will deliver the complete dual-track solution required for long-term ecological restoration and the engineered molecular erasure of synthetic contaminants from the biosphere.

References

1. Alimba, C. G., & Faggio, C. (2019). Microplastics in the marine environment: A review of the sources, occurrence, fate, and effects. Environmental Toxicology and Pharmacology, 72, 103251.

2. Chan, C. M., et al. (2019). Polyhydroxyalkanoates (PHAs) as sustainable alternatives to synthetics. European Polymer Journal, 116, 406–419.

3. Rochman, C. M., et al. (2013). Policy: Classify plastic waste as hazardous. Nature, 494(7436), 169–171.

4. Yoshida, S., et al. (2016). A bacterium that degrades and assimilates poly(ethylene terephthalate). Science, 351(6278), 1196–1199.

5. Wei, R., & Zimmermann, W. (2017). Microbial enzymes for the recycling of recalcitrant petroleum-based plastics: How far are we? Microbial Biotechnology, 10(6), 1308–1322.

6. Tournier, V., et al. (2020). An engineered PET depolymerase to break down and recycle plastic bottles. Nature, 580, 216–219. (Context for advanced enzymatic systems).

7. Gayathri, P.V., et al. (2023). Advanced oxidation processes for the degradation of microplastics: A review. Water Environment Journal.

8. Bautista-Zamudio, P.A., et al. (2023). Biodegradation of plastics by white-rot fungi: A review. Science of the Total Environment.

9. Long, B., et al. (2025). Remediation and upcycling of microplastics by algae with wastewater nutrient removal and bioproduction potential. Nature Communications, 16, 11570.

10. Rahimi Fard, D. (2026). The Synthetic Biosphere: From Petrochemical Invasion to Engineered Molecular Erasure. Amazon KDP.

Chapter 10: Closed-Loop Molecular Recycling, Zero-Shedding Design, and Global Governance – Industrial Circularity and Policy Frameworks

Introduction

Chapter 9 addressed the critical downstream imperative of active remediation — capturing and mineralizing legacy microplastic pollution already present in aquatic and marine environments through physical separation, advanced oxidation processes, and biological degradation pathways. While these technologies are essential for ecological restoration, they treat the symptoms of synthetic contamination rather than its root cause.

This chapter completes the solutions arc by focusing on upstream prevention and industrial circularity: the redesign of textile manufacturing, consumption, and end-of-life systems to eliminate emissions at source and close material loops. It integrates the material foundations of Chapter 7 (SSbD, biopolymers, finishing), the digital and recycling frameworks introduced in Chapter 8, and the remediation capabilities of Chapter 9 of this monograph into a unified, closed-loop strategy for a zero-shedding textile economy.

The industrial circularity and governance framework detailed here encompasses:

- The full spectrum of textile recycling methodologies (mechanical, thermomechanical, and chemical/molecular depolymerization) and the challenge of secondary microplastic generation within recycling processes themselves.
- Technical barriers posed by multi-fiber blends, chemical additives, and separation challenges, along with advanced solvent, enzymatic, and hybrid solutions.

- Global policy instruments — including modulated Extended Producer Responsibility (EPR), Digital Product Passports (DPPs), ecodesign mandates, and recycled content targets — required to align economic incentives with circular design.
- International governance, trade policy, standards harmonization, and just transition considerations essential for scaling circular systems across global value chains.

10.1 The Industrial Circular Loop: Textile Recycling Methodologies

To permanently prevent synthetic polymers from entering the biospheric sink, the global textile industry must replace linear consumption models with closed-loop recycling systems. These recycling technologies are divided into three main processing types: mechanical, thermomechanical, and chemical. The choice of pathway has profound implications for material quality, energy use, and secondary emissions.

10.1.1 Mechanical Recycling

Mechanical recycling is the most common recycling pathway, particularly for cotton and polyester-cotton blends. Post-consumer garments are sorted by color, stripped of trims (zippers, buttons), and fed through rotating tearing cylinders lined with metal pins. This process mechanically tears and shreds the fabric back into loose, staple fiber bundles. Because this tearing process physically breaks and shortens the fibers, mechanical recycling typically downcycles the materials into lower-value products like insulation, carpet padding, or wipes, rather than high-performance apparel yarns. Fiber shortening and contamination limit closed-loop applications.

10.1.2 Thermomechanical Recycling

Thermomechanical recycling is primarily used for clean, single-polymer synthetic textile waste (such as pure PET or Nylon). The material is shredded, washed, dried, and fed into a twin-screw extruder. Here, the polymer is heated above its melting point

(≈260°C for PET) and extruded through a die into continuous strands, which are cooled in a water bath and chopped into pellets. While highly efficient, each thermomechanical recycling cycle subjects the polymer to high thermal and mechanical shear, driving polymer degradation and shortening chain lengths. The number of viable recycling cycles is limited by cumulative thermal degradation; stabilizers and chain extenders can partially mitigate this.

10.1.3 Chemical Recycling (Molecular Depolymerization)

Chemical recycling uses chemical solvents, catalysts, and heat to break the ester or amide bonds of synthetic polymers, converting them back into their monomer precursors. This pathway offers the highest material value recovery and is the only technology capable of handling blended and contaminated textile waste.

- PET Glycolysis: Treats polyester waste with excess ethylene glycol (EG) in the presence of transesterification catalysts at temperatures between 180–250°C to produce bis(2-hydroxyethyl) terephthalate (BHET) monomers. The purified BHET can be re-polymerized into virgin-quality polyester.
- PET Methanolysis: Uses high-pressure methanol vapor at 180–220°C to depolymerize PET into dimethyl terephthalate (DMT) and ethylene glycol (EG).
- Polyamide (Nylon 6) Hydrolysis: Nylon 6 is depolymerized using superheated steam at 250–300°C and phosphoric acid catalysts to yield pure ε-caprolactam monomers.
- Pyrolysis: For mixed-plastic and multi-fiber textile waste, pyrolysis thermally cracks the materials in an oxygen-free reactor at temperatures between 400–600°C, converting the polymers into pyrolysis oil, syngas, and solid char that can be refined into petrochemical feedstocks.

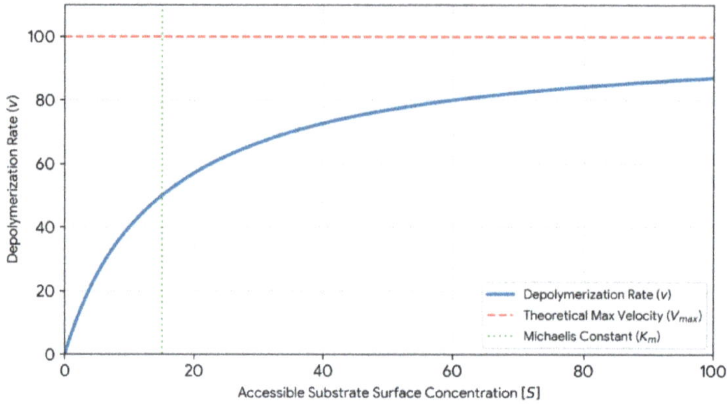

Figure 10.1: Operational Thermal Boundaries for Chemical Recycling Pathways

10.1.4 Process Intensification and Economic Viability of Chemical Recycling

The economic viability of chemical recycling at industrial scale hinges on process intensification strategies that reduce capital and operating expenditures while maximizing monomer yield and purity. Key intensification pathways include continuous-flow reactor designs (reducing reactor volume by 50–80%), integrated solvent and catalyst recovery loops (>99% recycling efficiency), reactive distillation and membrane separation for in-situ product removal, and modular, distributed recycling facilities co-located with collection hubs. Techno-economic analyses indicate that for PET chemical recycling, monomer production costs can approach or undercut virgin petrochemical routes when plant capacity exceeds 50,000 tonnes/year and feedstock collection costs are below $200/tonne. Achieving these thresholds requires coordinated investment in collection infrastructure, standardized sorting (enabled by DPPs), and policy support such as carbon pricing or recycled content mandates.

10.2 Microplastic Attrition Dynamics inside the Recycling Loop

While recycling reduces solid waste, the recycling process itself can generate significant volumes of microplastics and microfibers, creating a secondary source of environmental emissions. Understanding and mitigating these internal emission hotspots is essential for ensuring that recycling systems do not inadvertently increase environmental loading.

10.2.1 Airborne Fiber Emissions during Mechanical Shredding

During mechanical shredding, garments are exposed to high-velocity mechanical tearing forces. The pin cylinders of tearing machines exert high shear and tensile stresses on the dry textiles, exceeding the yield strength of individual fibers and causing physical fractures. This mechanical fatigue generates a high concentration of short, respirable airborne microfibers (typically <100 μm in length), which can escape into the factory air. High-capacity localized dust extraction and HEPA filtration systems are essential at tearing and carding stations to protect workers and prevent environmental release.

10.2.2 Waterborne Microplastic Discharges from Washing and Separation

Before thermomechanical or chemical recycling, post-consumer polyester and polyamide garments must be washed to remove surface contaminants, labels, and adhesives. This washing process uses heavy-duty rotary wash drums, hot water, and surfactants, creating a high-energy wet environment. The mechanical friction between garments and the drum walls, combined with surfactant-induced fiber swelling, triggers the release of millions of microfibers. High-efficiency, multi-stage filtration or membrane bioreactor systems on wastewater discharge lines are required to prevent large quantities of microplastics from entering municipal sewer systems.

10.2.3 The Altered Shedding Behavior of Recycled Yarns

Recycled synthetic yarns—particularly those derived from mechanical or thermomechanical recycling—often exhibit significantly higher shedding rates than their virgin counterparts during subsequent consumer use. This increased shedding propensity is driven by several structural changes: chain-scission degradation during thermomechanical extrusion (shortening molecular chain length and increasing polydispersity), surface decortication and micro-cracks (which propagate during consumer use), and shortened fiber lengths (which reduce binding forces within the yarn core). Advanced surface coatings and high-twist yarn designs on recycled textile products are therefore essential to mitigate these effects.

10.3 Multi-Fiber Blends, Chemical Additives, and Separation Barriers

A key barrier to scaling circular closed-loop recycling is the high complexity of post-consumer textiles, which are rarely composed of a single, pure polymer. The intimate blending of different fiber types, combined with the presence of chemical additives and dyes, creates significant technical and economic challenges for material recovery.

10.3.1 The "Intimate Blend" Separation Challenge

Many garments are constructed using blended yarns—such as polycotton (polyester-cotton) or poly-elastane (polyester-spandex)—where different fiber types are intimately spun together. Separating these blends requires advanced solvent systems that can selectively dissolve one fiber type while leaving the other intact. Ionic liquids like 1-butyl-3-methylimidazolium chloride ([BMIM]Cl) can dissolve cellulose pulp from polycotton blends, allowing the cotton fraction to be recovered as regenerated cellulose fibers (Lyocell), while the polyester fraction remains solid and can be collected for chemical recycling. Enzymatic approaches (e.g., cellulases) offer a milder alternative for selective digestion of natural fiber fractions.

10.3.2 The Problem of Chemical Additives and Plasticizers

During textile manufacturing, raw polymers are blended with various chemical additives to improve flexibility, color, and performance: plasticizers and flame retardants (phthalates, organophosphate flame retardants, PFAS), and heavy-metal catalysts (antimony trioxide). During mechanical and thermomechanical recycling, these additives remain in the recycled polymer, causing color defects, chemical degradation, and potential toxicity issues. Chemical recycling (depolymerization) bypasses these issues by breaking the polymer down into its monomers, allowing additives, dyes, and catalysts to be separated and removed during monomer purification, yielding clean, virgin-quality polymer.

10.4 Global Policy Directives, Extended Producer Responsibility (EPR), and Digital Product Passports

Transitioning to a circular textile economy cannot rely on voluntary industry action alone. It requires robust regulatory frameworks, standardized metrics, and transparent data infrastructures to align economic incentives with environmental safety.

10.4.1 Extended Producer Responsibility (EPR) Schemes

Extended Producer Responsibility (EPR) laws require textile brands and manufacturers to fund the collection, sorting, and recycling of their products at the end of their useful life. To incentivize sustainable design, EPR fees can be modulated based on the garment's environmental impact. Brands that design garments using single-polymer fibers, continuous filament yarns, and non-toxic dyes pay lower fees, whereas brands that produce complex, high-shedding multi-fiber blends containing hazardous additives pay significantly higher fees. This fee modulation provides a direct financial incentive for brands to adopt Safe and Sustainable by Design (SSbD) principles.

10.4.2 Digital Product Passports (DPPs)

The European Union's Ecodesign for Sustainable Products Regulation (ESPR) mandates the introduction of Digital Product Passports (DPPs) for all textile products. A DPP is a digital record accessible via a QR code or RFID tag embedded in the garment. It contains comprehensive data on the product's material composition, chemical additives, supply chain carbon footprint, and disassembly instructions. At end-of-life sorting facilities, high-speed automated sorting systems can scan the DPP to immediately identify the garment's fiber composition, routing it to the appropriate mechanical or chemical recycling stream and preventing contamination.

10.4.3 Ecodesign and Recycling Mandates

Regulatory frameworks are increasingly setting minimum targets for recycled content in new textile products, creating a reliable market demand for recycled polymers and driving investment in advanced chemical recycling infrastructure. Additionally, new regulations prohibiting the landfilling or destruction of unsold garments are forcing brands to optimize their production volumes and adopt circular take-back systems, helping to decouple industry growth from primary resource consumption.

10.5 Truly Closed-Loop, Zero-Shedding Circular Systems: A Strategic Synthesis

Developing a truly circular, zero-shedding textile system requires aligning upstream material selection, midstream consumer care, and downstream end-of-life processing into a single, coordinated framework.

To implement this strategic synthesis, the global textile industry must execute three key interventions:

1. Design for Disassembly and Recycling: Garments should be designed from the start to be easily recycled, using single-polymer constructions, soluble sewing threads, and removable trims to

simplify sorting and processing. This upstream intervention is the most cost-effective way to enable circularity.

2. Close the Chemical Loop: Brands must invest in and scale advanced chemical and enzymatic depolymerization technologies, converting post-consumer textile waste into high-purity monomers that can replace primary petrochemical feedstocks. This downstream intervention is essential for achieving true material circularity.

3. Transition to Biodegradable Alternatives: For applications where fiber shedding cannot be prevented (such as technical outerwear or workwear), traditional petroleum-based synthetic polymers should be replaced with marine-degradable biopolymers like PHAs or high-performance cellulosics like Lyocell. This material substitution intervention addresses the fundamental persistence of synthetic debris in the environment.

10.6 Global Governance, Trade Policy, and the Geopolitics of Circular Textile Systems

The transition to a truly circular, zero-shedding textile economy is not merely a technical or national policy challenge — it is a profoundly geopolitical one. Global textile supply chains span dozens of countries, with production concentrated in a handful of manufacturing hubs (China, India, Bangladesh, Vietnam, Turkey) while consumption and regulatory leadership remain concentrated in Europe, North America, and East Asia.

10.6.1 Trade Policy as a Lever for Circular Design

Trade policy instruments — tariffs, non-tariff barriers, rules of origin, and preferential trade agreements — can be strategically deployed to accelerate the adoption of circular design principles. Tariff reductions or duty-free access could be conditioned on compliance with minimum recycled content thresholds, verified low-shedding performance, or participation in certified take-back and recycling schemes. Conversely, higher tariffs or import restrictions could be applied to high-shedding, difficult-to-recycle textile products. Rules of origin in free trade agreements could be reformed to recognize recycled content and remanufactured goods, facilitating cross-border flows of secondary raw materials.

10.6.2 International Standards Harmonization and Mutual Recognition

The proliferation of national and regional standards for microplastic emissions, biodegradability claims, recycled content verification, and Digital Product Passport formats creates significant compliance costs and market fragmentation. International harmonization of testing methods (building on ISO and AATCC protocols), mutual recognition agreements for certification schemes, and interoperable DPP data architectures are essential for reducing barriers to trade in circular textile products and for enabling efficient global recycling markets. Bodies such as the International Organization for Standardization (ISO), the American Association of Textile Chemists and Colorists (AATCC), the Global Fashion Agenda, and the Ellen MacArthur Foundation's Circular Fibres Initiative play catalytic roles in developing globally accepted standards.

10.6.3 Addressing Distributional Impacts and Just Transition

The shift from linear to circular textile systems will create winners and losers across the global value chain. Countries and regions heavily dependent on virgin polymer production, conventional mechanical recycling (downcycling), or low-value textile manufacturing may face economic disruption, while those positioned to lead in chemical recycling, bio-based materials, and digital traceability stand to gain. A just transition framework — analogous to those emerging in the energy sector — must be integrated into global circular textile governance. This includes financial and technical support for developing countries to build advanced sorting and recycling infrastructure, preferential market access for circular textile products from least-developed countries, reskilling and social protection programs for workers displaced from linear production models, and mechanisms to ensure that the environmental benefits of circularity are equitably shared.

10.6.4 Multilateral Environmental Agreements and the Global Plastics Treaty

The ongoing negotiations toward a legally binding global plastics treaty under the United Nations Environment Assembly represent a once-in-a-generation opportunity to establish binding international obligations for microplastic emission reduction, design standards for high-shedding textile categories, and harmonized extended producer responsibility frameworks. Provisions addressing textile microfibers — such as mandatory microfiber filters on new washing machines, design-for-recyclability requirements, and targets for recycled content — could be integrated into the treaty text or its protocols. Linkages with existing multilateral environmental agreements (the Basel Convention, the Stockholm Convention, and the Convention on Biological Diversity) can create a coherent international legal architecture for addressing the full lifecycle of synthetic textiles.

Synthesis and Conclusion: The Restorative Path Forward

This chapter has detailed the midstream-to-downstream and governance interventions required to close the textile material loop—industrial recycling methodologies (with attention to secondary emissions), solutions to multi-fiber and additive barriers, modulated EPR, Digital Product Passports, ecodesign mandates, and international governance frameworks including trade policy and just transition measures. These complete the upstream prevention and circularity vision initiated in Chapters 7–8 and complement the active remediation strategies of Chapter 9. Together, the integrated upstream-to-downstream framework presented across Chapters 7–10 delivers the complete dual-track strategy needed to eliminate synthetic textile emissions from the global biosphere and advance the engineered molecular erasure of contaminants from the biosphere.

References

1. AATCC Committee RA100. (2021). AATCC Test Method 127-2021: Test Method for Microfiber Release During Home Laundering. Research Triangle Park, NC.

2. Kang, H., et al. (2021). Impact of Chitosan Pretreatment to Reduce Microfibers Released from Synthetic Textiles during Laundering. Water, 13(18), 2480.

3. Niinimäki, K., et al. (2020). The environmental price of fast fashion. Nature Reviews Earth & Environment, 1(4), 189–200.

4. European Commission (2022/2024). Ecodesign for Sustainable Products Regulation (ESPR) and Digital Product Passport framework. Brussels.

5. Wei, R., & Zimmermann, W. (2017). Microbial enzymes for the recycling of recalcitrant petroleum-based plastics: How far are we? Microbial Biotechnology, 10(6), 1308–1322.

6. Tournier, V., et al. (2020). An engineered PET depolymerase to break down and recycle plastic bottles. Nature, 580, 216–219.

7. OECD (2023). New Aspects of EPR: Extending producer responsibility to additional product groups and environmental impacts. OECD Environment Working Papers No. 217.

Chapter 11: The Consumption Vector – Fast Fashion, Behavioral Economics, and the Demand-Side Arithmetic of Microplastic Emissions

Introduction

Chapters 7 through 10 have constructed a technically coherent supply-side framework for reducing textile-derived microplastic emissions. This framework encompasses Safe and Sustainable by Design (SSbD) material engineering, continuous-filament yarn architectures, advanced surface coatings, digital traceability systems, molecular recycling infrastructure, modulated Extended Producer Responsibility (EPR), and downstream remediation technologies. These interventions target the intrinsic shedding propensity of textile materials and the technical systems required to capture or degrade emissions after release.

This supply-side architecture, while necessary, is strategically incomplete. The dominant driver of cumulative microplastic loading from textiles is not the shedding rate per individual garment. It is the total volume of garments that enter the use phase, are laundered repeatedly, and are eventually discarded. Fast fashion's structural acceleration of production and consumption cycles has multiplied garment throughput by a factor of several times relative to historical patterns (Ellen MacArthur Foundation, 2017; Niinimäki et al., 2020). No incremental improvement in per-garment shedding performance can fully offset this multiplication of units. A 60% reduction in microfiber release per wash is nullified if the number of garments washed annually increases by 150%.

This chapter addresses the demand-side arithmetic of microplastic emissions. It examines the quantitative relationship between consumption velocity and cumulative microfiber release across garment lifecycles, the behavioral economics that sustain accelerated replacement, the potential of alternative consumption models to reduce throughput, and the information architectures required to support lower-impact consumer decisions. By integrating demand-side interventions with the supply-side foundations established in Chapters 7–10, the monograph completes the dual-track strategy required for systemic reduction of textile-derived microplastic pollution.

11.1 The Arithmetic of Throughput: Consumption Velocity and Cumulative Microfiber Emissions

The cumulative microfiber emissions attributable to a garment are a function of two independent variables: the mass of fibers shed per wash cycle (determined by fiber type, yarn construction, fabric structure, and finishing treatments) and the total number of wash cycles the garment experiences before disposal. Supply-side interventions primarily address the first variable. Demand-side interventions address the second.

Consider a representative polyester T-shirt. Under standardized domestic laundering conditions, such a garment sheds approximately 0.3–0.7 mg of microfibers per wash cycle. If the garment remains in active use for 50 wash cycles before disposal, total microfiber release is 15–35 mg. If fast-fashion economics compress active lifespan to 10 wash cycles, total release per garment falls to 3–7 mg. However, if the consumer purchases and discards five such garments per year instead of one, the manufacturing emissions associated with this garment type increase fivefold. While use-phase (laundering) emissions may be comparable when total wash cycles are held constant, the pre-consumer manufacturing burden — energy, water, chemicals, and upstream microplastic losses during production — is multiplied. When both manufacturing and laundering phases are considered, annual microfiber-related loading rises substantially.

This throughput multiplier is not hypothetical. Global apparel production has approximately doubled since 2000, with the fast-fashion segment accounting for a disproportionate share of growth. The compression of design-to-retail cycles from 12–18 months to 2–6 weeks, and the expansion from two seasonal collections to 20–52 micro-seasons, has structurally increased the number of garments entering the use phase annually. Each additional garment contributes its full manufacturing footprint and its full complement of laundering emissions, regardless of incremental improvements in fiber-level shedding performance.

Empirical studies of clothing longevity confirm the scale of under-utilization. Research in the United Kingdom and Nordic countries indicates that a significant proportion of fast-fashion garments are worn fewer than ten times before disposal. This pattern represents both an economic inefficiency and an environmental multiplier: the manufacturing and laundering emissions associated with each garment are amortized over far fewer uses than the garment's technical durability would permit.

The policy implication is direct. Supply-side interventions that achieve a 50–70% reduction in shedding per wash (realistic through optimized yarn twist, continuous-filament constructions, and durable surface coatings) can be entirely offset by a doubling of garment throughput. Demand-side strategies that extend active garment lifespan or reduce the number of new garments entering circulation therefore offer leverage comparable to, and potentially greater than, technical improvements in material design.

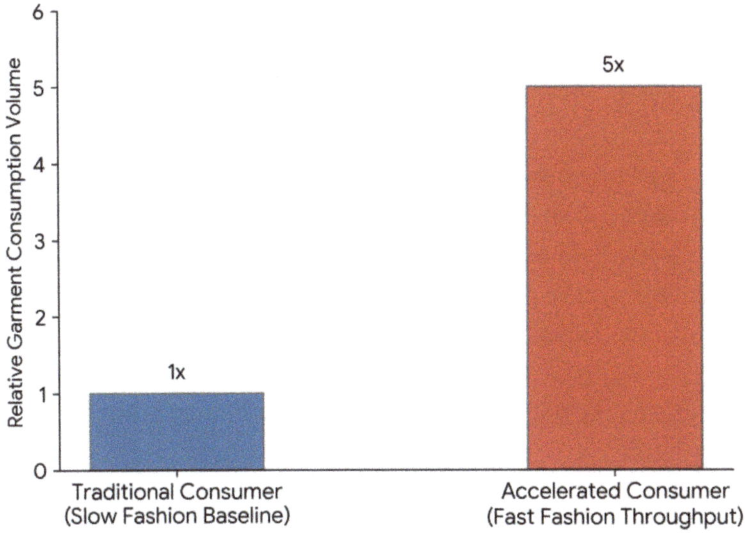

Figure 11.1: Linear Garment Throughput Multiplier Effect

11.2 Behavioral Economics of Perceived Obsolescence and Accelerated Replacement Cycles

Fast fashion's economic model is predicated on the deliberate acceleration of perceived obsolescence. This acceleration is achieved through low price points that reduce the financial and psychological cost of disposal, rapid trend cycling that renders garments socially outdated before physical failure, and marketing systems that frame clothing acquisition as a form of identity refreshment rather than functional replacement.

Behavioral economics identifies multiple mechanisms that reinforce this pattern. The endowment effect is attenuated when acquisition costs are low; consumers exhibit less reluctance to discard items acquired at minimal financial or psychological cost. Present bias leads individuals to overweight immediate gratification from new purchases while discounting the cumulative environmental and financial costs of frequent replacement. Social signaling dynamics amplify throughput: in contexts where clothing functions as a visible marker of status, group affiliation, or currency within attention economies, the pressure to update wardrobes intensifies as trend velocity increases.

Research on clothing use patterns demonstrates that the average number of wears per garment has declined substantially over the past two decades. This decline is most pronounced in the fast-fashion segment, where garments are frequently designed and marketed with the expectation of short active lifespans. The resulting under-utilization constitutes a structural driver of microfiber emissions that cannot be addressed through fiber-level engineering alone.

Effective demand-side intervention must therefore target the structural incentives that make frequent replacement the default rational choice. These incentives include price signals that fail to internalize end-of-life costs, information asymmetries that obscure durability and shedding performance, and social norms that equate frequent acquisition with status or self-expression. Altering these incentives requires coordinated action across economic instruments, information architecture, and cultural framing.

11.3 Alternative Consumption Models as Throughput Reduction Strategies

Models that decouple garment use from ownership offer direct mechanisms for reducing throughput while preserving consumer access to variety and novelty. These models operate on different economic logics than traditional ownership and create different incentives for durability and utilization.

Rental and subscription systems convert clothing from a purchased asset into a managed service. The provider's economic interest shifts from maximizing unit sales to maximizing the number of wears extracted from each garment. This alignment creates a commercial rationale for durability, repairability, and low-shedding design — attributes that directly reduce microplastic emissions. A garment circulating in a well-managed rental fleet may experience 50–100 wear cycles rather than the 5–10 cycles typical of fast-fashion ownership models, spreading manufacturing and laundering emissions across a substantially larger number of uses.

Peer-to-peer and brand-operated resale platforms extend garment lifespan by creating secondary markets. Research indicates that garments entering resale streams are typically worn two to three times longer than garments that proceed directly to disposal. The behavioral mechanism is partly economic: the possibility of recovering residual value at end of use reduces the net cost of ownership and creates an incentive for consumers to select higher-quality, longer-lasting items.

These models are not automatically environmentally superior. Rental systems can increase transportation emissions and require industrial-scale cleaning that generates additional microfiber releases if filtration is inadequate. Resale markets can expand overall consumption if they function as a supplement rather than a substitute for new production. The net benefit depends on implementation details — particularly whether the alternative model demonstrably displaces new garment production and whether cleaning and logistics systems incorporate microfiber capture technologies.

Policy can strengthen the environmental performance of these models. Requirements for microfiber filtration on commercial rental cleaning systems, fiscal incentives for repair and resale enterprises, and mandates for brand participation in certified take-back and resale programs can amplify the throughput-reduction potential of alternative consumption architectures while minimizing unintended emission increases.

11.4 Consumer Information Architecture: Digital Product Passports and Durability Labeling

Information interventions are most effective when they reduce cognitive friction at the moment of decision. Current textile labeling primarily communicates fiber composition and care instructions. It rarely conveys the variables most relevant to cumulative microplastic emissions: expected durability, shedding performance under standardized testing, or repairability.

Digital Product Passports (DPPs), required under the European Union's Ecodesign for Sustainable Products Regulation, create an opportunity to reorient consumer information toward variables that influence microfiber loading. When designed as consumer-facing interfaces rather than purely regulatory compliance mechanisms, DPPs can communicate:

- Verified microfiber shedding performance under standardized test methods
- Expected number of wear or wash cycles before significant degradation
- Repairability score and availability of repair services or spare parts
- Circularity credentials, including recycled content and participation in take-back or resale programs

Behavioral research on eco-labeling indicates that simple, comparative metrics are more effective at influencing purchasing decisions than complex technical specifications. DPPs can be structured to surface comparative information (for example, "this garment ranks in the top quartile for durability within its category") at the point of sale through QR codes or NFC interfaces, reducing the cognitive effort required for consumers to identify lower-impact options.

Information architecture is most powerful when paired with economic incentives. A consumer who understands that a garment sheds heavily or has low expected durability may still purchase it if the price differential is large and social pressure to update is strong. Information interventions therefore achieve greater impact when combined with modulated EPR fees that reward durability and low-shedding performance, and with alternative consumption models that make lower-throughput choices more convenient or economically attractive.

11.5 Behavioral Change Research and Multi-Level Intervention Design

Psychological and sociological research on sustainable consumption consistently demonstrates that individual-level interventions — information provision, eco-labeling, and value-based appeals — produce modest and often transient behavioral change when structural conditions remain unchanged. Fast fashion's combination of low prices, high convenience, and powerful social signaling creates a choice architecture in which frequent replacement is the default option.

Effective demand-side strategies must therefore operate across multiple levels simultaneously:

At the individual level, information and labeling can reduce cognitive friction for lower-impact choices when presented at the point of decision.

At the social level, normalization of repair, resale, and extended garment use can shift perceived norms and reduce the social cost of slower consumption patterns.

At the structural level, economic instruments (EPR fee modulation based on durability and shedding performance, tax treatment of repair services, restrictions on destruction of unsold goods) alter the relative cost of high- versus low-throughput consumption.

At the cultural level, reframing of fashion from rapid trend consumption toward curation, care, and longevity can reshape the meaning attached to clothing acquisition and use.

The most effective interventions integrate multiple levels. A brand that offers durable, low-shedding garments supported by transparent DPP information, a robust repair and resale program, and take-back incentives creates a choice architecture in which the lower-impact option is simultaneously more economically rational and socially supported. Policy can accelerate this integration by mandating minimum durability standards for certain garment categories, requiring verified lifespan and shedding data in DPPs, and creating fiscal incentives that reward business models aligned with reduced throughput.

Synthesis: Integrating Demand-Side and Supply-Side Strategies

The demand-side analysis presented in this chapter does not replace the supply-side interventions developed in Chapters 7–10. It complements and multiplies their effect. A garment engineered for low shedding that is worn and washed only ten times before disposal still contributes less to cumulative emissions than a high-shedding garment worn fifty times. The greatest reduction, however, is achieved when both conditions are satisfied: low intrinsic shedding propensity and extended active lifespan under conditions of slower throughput.

The integrated strategy therefore requires simultaneous action on both sides of the emissions equation. Supply-side interventions reduce the mass of microfibers released per wash cycle through material redesign, surface engineering, and filtration technologies. Demand-side interventions reduce the total number of wash cycles across the system by extending garment lifespan, enabling alternative consumption models, and reshaping the economic and informational conditions that currently favor high throughput.

Chapters 7–10 provide the technical toolkit for the supply side. This chapter provides the behavioral, economic, and informational analysis for the demand side. Together they constitute the complete dual-track framework required to address both the material properties of textiles and the systemic drivers of their over-production and under-utilization.

The ultimate objective is not merely to engineer garments that shed less, but to engineer a textile economy in which the default consumption pattern — slower, more intentional, and higher-utilization — aligns material performance, economic incentives, and planetary boundaries. Achieving this alignment requires that demand-side strategies receive analytical and policy attention commensurate with the supply-side innovations that have dominated the discourse on textile sustainability to date.

References

1. Ellen MacArthur Foundation. (2017). A New Textiles Economy: Redesigning Fashion's Future. Cowes, UK.

2. Niinimäki, K., et al. (2020). The environmental price of fast fashion. Nature Reviews Earth & Environment, 1(4), 189–200.

3. European Commission. (2022/2024). Ecodesign for Sustainable Products Regulation (ESPR) and Digital Product Passport framework. Brussels.

4. Sandin, G., & Peters, G. M. (2018). Environmental impact of textile reuse and recycling – A review. Journal of Cleaner Production, 184, 353–365.

5. Laitala, K., & Klepp, I. G. (2020). Clothing longevity: The role of consumer behaviour and garment design. Sustainable Production and Consumption, 24, 1–12.

6. OECD. (2023). New Aspects of EPR: Extending producer responsibility to additional product groups and environmental impacts. OECD Environment Working Papers No. 217.

7. United Nations Environment Programme (UNEP). (2023). Turning Off the Tap: How the world can end plastic pollution and create a circular economy. Nairobi.

8. Fletcher, K. (2016). Craft of Use: Post-Growth Fashion. Routledge.

9. ThredUp. (2024). Resale Report 2024. https://www.thredup.com/resale-report (Accessed June 2026).

10. Rent the Runway. (2023). Impact Report: Measuring the Environmental Benefits of Rental. New York.

Chapter 12: The Predictive Turn — Multi-Scale Modeling, Machine Learning, and the Computational Future of Textile Microplastic Science

Introduction

The science assembled in the preceding eleven chapters is, with few exceptions, descriptive. It catalogues the sources of textile microfibers, measures the mass they shed, characterizes the particles that result, traces their movement through environmental compartments, and documents their biological effects. This is the necessary first phase of any environmental science: one must be able to observe and quantify a phenomenon before one can act on it. But description has a structural limit. It is retrospective. It tells us what a fabric has shed, after the fabric exists and after the shedding has occurred. It cannot tell us what a fabric not yet manufactured will shed, nor what aggregate burden a production decision made today will impose on a watershed a decade hence.

The frontier that distinguishes the next phase of this field is predictive rather than descriptive. A predictive capacity would allow a designer to screen a candidate fabric in silico before a single specimen is woven, ranking construction and finish options by their forecast shedding behavior. It would allow a regulator to estimate the emission consequences of a policy before it is enacted. It would allow a life-cycle assessment practitioner to estimate the microfiber burden of a product line without physically testing every garment. None of this is possible with measurement alone. It requires models that map from causes to consequences, calibrated against comparable data and honest about their uncertainty.

This monograph has, in the course of its argument, introduced three quantitative constructs that have so far been presented in isolation: the two-phase kinetic emission model of Chapter 3, the Yarn Shedding Index of Chapter 7, and the surface-mediated enzymatic depolymerization model of Chapter 8. This chapter advances a proposition: that these constructs are not three unrelated formulas but three components of a single, multi-scale predictive framework, and that the gaps between them can in principle be bridged by data-driven machine-learning methods of the kind introduced in Chapter 6. What follows is explicitly a proposed architecture, not a validated tool. Its purpose is to define the computational program that would move textile microplastic science from documentation toward design.

12.1 From Description to Prediction: The Computational Gap

The published literature on textile microfiber emissions, including the survey treatments that dominate the reference shelf, is overwhelmingly empirical and retrospective. Studies report shedding values for specific fabrics under specific protocols; reviews aggregate those values into ranges. This body of work is indispensable, but it does not constitute a predictive science, because it provides no general function that maps from a fabric's measurable properties to its expected emission behavior. Each new fabric must be tested anew. There is no equivalent, in this field, of the quantitative structure-property relationships that allow chemists to estimate the properties of an untested molecule from its structure.

The computational gap is therefore the absence of a transfer function: a relationship, mechanistic or statistical or both, that takes as input the measurable descriptors of a textile — fiber polymer, staple length, yarn twist, fabric construction, mass per unit area, finish chemistry — and returns as output a forecast of its fragmentation behavior across its life cycle. Closing this gap is the precondition for prevention by design, which Chapter 7 argued is the only intervention with leverage proportional to the scale of the problem. One cannot design for low shedding if one cannot predict shedding. The remainder of this chapter sets out how the components already developed in this monograph might be assembled into such a transfer function.

12.2 The Component Models: A Recap and Reframing

The framework proposed here rests on three constructs introduced earlier, which are now reframed as components operating at different scales and stages of the emission pathway.

The Yarn Shedding Index, introduced in Section 7.2.1, is a dimensionless construct that estimates a fabric's intrinsic shedding propensity from its structural parameters — fiber type, staple configuration, yarn twist, and construction. It operates at the material scale, before any wash event, and it is in principle computable from specification data alone. In the architecture proposed here it functions as the entry point: a forecast of propensity derived from design parameters.

The two-phase kinetic emission model, introduced in Section 3.1.2, describes the mass of fragments released over the course of a laundering cycle as the sum of a rapidly decaying transient term, representing the release of pre-existing loosely bound surface fragments, and an ongoing generation term, representing fragments produced by mechanical and chemical action during the cycle. It operates at the event scale, converting propensity into a time-resolved emission per wash. In the proposed architecture it is the second stage: it translates a structural propensity into an emitted mass under specified wash conditions.

The surface-mediated enzymatic depolymerization model, introduced in Section 8.3.2, describes the rate at which released polymer is broken down under enzymatic action as a function of accessible substrate surface area. It operates at the fate-and-remediation scale, downstream of emission. In the proposed architecture it represents the terminal stage: the post-release transformation of emitted material.

Individually, each model is descriptive of its own scale. The proposition of this chapter is that, chained together and supplied with the consumption-velocity term developed in Chapter 11, they describe the full pathway from a design decision to an environmental flux.

12.3 A Multi-Scale Architecture

The proposed framework links four scales in sequence, with the output of each scale serving as an input to the next.

At the material scale, design descriptors are mapped to an intrinsic shedding propensity via the Yarn Shedding Index. At the event scale, that propensity is combined with wash conditions — temperature, mechanical load, detergent presence, cycle duration — through the two-phase kinetic model to yield an emitted mass per laundering cycle. At the product-lifetime scale, the per-cycle emission is integrated over the number of wash cycles a garment undergoes before disposal, a quantity governed by the consumption-velocity dynamics analyzed in Chapter 11; this is the point at which the demand-side multiplier enters the calculation, since a garment washed ten times before discard and one washed fifty times impose very different lifetime burdens even at identical per-cycle emission. At the population scale, product-lifetime emissions are aggregated across the garments entering a market to yield a regional emission inventory, which then becomes the source term for the environmental transport and partitioning models discussed in Chapter 2.

Table 12.1 summarizes the framework's components, the transfer function operating at each scale, and the calibration status of each stage.

Table 12.1: The Multi-Scale Predictive Framework — Components, Transfer Functions, and Calibration Status

Scale	Input	Model / Method (source)	Output	Calibration Status
Material	Fabric design descriptors (fiber type, yarn twist, construction, finish)	Yarn Shedding Index, S_f (Section 7.2.1)	Intrinsic shedding propensity	Functional form specified; coefficients require empirical calibration
Event (per wash cycle)	Shedding propensity plus wash conditions (temperature, mechanical load, detergent, duration)	Two-phase kinetic emission model (Section 3.1.2)	Emitted mass per laundering cycle	Functional form specified; coefficients require empirical calibration
Product lifetime	Per-cycle emission plus wash-count distribution	Integration over wash cycles, with consumption-velocity term (Chapter 11)	Lifetime emission per garment	Conceptual; requires consumption and utilization data
Population / environmental	Aggregated lifetime emissions across the garments entering a market	Emission inventory coupled to transport and partitioning (Chapter 2)	Regional environmental flux (source term)	Conceptual; requires aggregation and transport coupling

Expressed as a conceptual chain, the architecture proceeds from design descriptors, through propensity, through per-cycle emission, through lifetime emission, to aggregate flux. Each arrow in this chain is a transfer function. Some of these functions are mechanistic and already specified in this monograph; others are not yet established and would require empirical calibration or data-driven estimation. The honest status of the framework is that it is a scaffold with three of its joints specified and the remainder awaiting data. Its value at present is conceptual: it identifies precisely which relationships must be quantified, and in what order, for the field to achieve predictive capacity.

12.4 The Machine-Learning Prediction Layer

Where mechanistic models are incomplete or where the underlying physics is too complex to specify from first principles, data-driven methods offer a complementary route. Chapter 6 established the use of machine learning for the classification of microplastic particles — determining what a given particle is. The predictive turn proposed here pivots that capability from classification to regression: from determining what a particle is, to forecasting how much a fabric will shed.

A supervised regression model for shedding prediction would take as features the measurable descriptors of a textile — polymer composition, yarn twist, staple length, fabric construction class, mass per unit area, and finish chemistry — and would be trained to predict a shedding mass under a specified test protocol. Tree-ensemble methods such as random forests and gradient-boosted trees are well suited to this kind of tabular, heterogeneous, moderately sized data, and the broader applicability of machine learning to microplastic problems is documented in the recent literature. Such a model would not replace the mechanistic framework of Section 12.3; it would fill the joints that mechanism cannot yet specify, and it would provide rapid in silico screening of candidate fabrics for the design application described in Chapter 7.

A more sophisticated and more distinctive approach is the physics-informed hybrid, in which a mechanistic model and a data-driven model are combined rather than used as alternatives. In this configuration the two-phase kinetic model supplies the known physical structure of the emission process, while a machine-learning component is trained to predict the residual — the systematic discrepancy between the mechanistic prediction and the measured value, which encodes the effects the mechanism omits. This strategy, established in computational science as physics-informed machine learning, preserves the interpretability and extrapolation behavior of the mechanistic model while capturing the unmodeled complexity that purely empirical fitting would otherwise be needed to represent. For a field with sparse, expensive, and protocol-fragmented data, the data efficiency of physics-informed methods is a decisive advantage.

The central obstacle to any of these approaches is not algorithmic but evidentiary. A machine-learning model is only as good as its training data, and the training data for shedding prediction is fundamentally compromised by the absence of a harmonized measurement standard. Shedding values in the published literature are generated under a range of incompatible test protocols — the principal methods, ISO 4484-1, AATCC TM212, and The Microfibre Consortium method, differ in whether detergent is included, in the number of specimens tested, and in the agitation device used — and these differences materially change the measured result. A model trained on a mixture of values generated under detergent and without it, on four-specimen and eight-specimen protocols, on different agitation devices, will learn the noise of the methods as readily as the signal of the fabrics. The predictive turn is therefore downstream of a harmonization imperative: comparable data is the raw material of prediction, and until measurement protocols converge, the data-driven layer of this framework will remain data-starved.

12.5 Uncertainty, Calibration, and the Honest Status of the Framework

A predictive framework that does not propagate its uncertainty is not a scientific instrument but a false promise. Each transfer function in the multi-scale architecture carries error, and because the functions are chained, those errors compound. An uncertainty in the Yarn Shedding Index propagates into the per-cycle emission estimate; an uncertainty in the wash-count distribution propagates into the lifetime emission; and the aggregate flux at the population scale inherits the accumulated uncertainty of every stage below it. A responsible implementation would carry uncertainty bounds through the entire chain and would report aggregate fluxes as distributions rather than point values. The compounding of error across scales is not a defect to be hidden; it is a property of the system that any honest forecast must represent.

The framework as presented here is uncalibrated. The mechanistic components have functional forms but their coefficients are not empirically fixed for the general case; the data-driven components are described but not trained, because the comparable data required to train them does not yet exist at sufficient scale. This is a deliberate and necessary acknowledgment. The contribution of this chapter is not a working predictor but a specification of the program required to build one: which relationships must be quantified, which data must be standardized and gathered, and how mechanistic and data-driven methods must be combined. A reader seeking a tool to deploy today will not find it here. A reader seeking to understand what the next decade of quantitative work in this field must accomplish will find it specified.

12.6 Research Frontiers and the Distinguishing Ambition

The predictive program sketched in this chapter opens several frontiers, each connected to an earlier argument of the monograph.

The first is in silico fabric screening, in which the material-scale layer of the framework is used to rank design options before manufacture, operationalizing the Safe and Sustainable by Design principle of Chapter 7 as a computational rather than a purely experimental practice. The second is the textile-lifecycle digital twin, a dynamic model of a garment or product line that integrates the full multi-scale chain and updates its emission forecasts as design, consumption, and end-of-life parameters change. The third is predictive policy modeling, in which the population-scale layer is used to forecast the emission consequences of regulatory interventions — filtration mandates, extended producer responsibility schemes, design standards — before they are enacted, giving the governance instruments of Chapter 10 a quantitative basis for ex ante evaluation rather than only ex post measurement.

These frontiers share a common character that defines the distinguishing ambition of this chapter and, by extension, of this monograph. The descriptive science that has dominated the field documents a problem. The predictive science proposed here would equip practitioners to design against it. The transition from the one to the other is the central intellectual movement that separates a computational treatment of textile microplastic pollution from a survey of it. The framework offered here is incomplete, uncalibrated, and dependent on a harmonized measurement base that does not yet fully exist. It is offered nonetheless, because naming the program precisely is the first step toward executing it, and because a field that can only describe what has already happened cannot design the future it needs.

References

1. Napper, I. E., & Thompson, R. C. (2016). Release of synthetic microplastic plastic fibres from domestic washing machines: Effects of fabric type and washing conditions. Marine Pollution Bulletin, 112(1–2), 39–45.
2. De Falco, F., Di Pace, E., Cocca, M., & Avella, M. (2020). The contribution of washing processes of synthetic clothes to microplastic pollution. Scientific Reports, 9, 6633.
3. Khanam, M. M., et al. (2025). Advances in machine learning for the detection and classification of microplastics. Frontiers in Environmental Science.
4. Raissi, M., Perdikaris, P., & Karniadakis, G. E. (2019). Physics-informed neural networks: A deep learning framework for solving forward and inverse problems involving nonlinear partial differential equations. Journal of Computational Physics, 378, 686–707.
5. ISO 4484-1:2023. Textiles and textile products — Microplastics from textile sources — Part 1: Determination of material loss from fabrics during washing. International Organization for Standardization, Geneva.
6. AATCC TM212-2021. Test Method for Fiber Fragment Release During Home Laundering. American Association of Textile Chemists and Colorists, Research Triangle Park, NC.
7. The Textile Matrix, Chapter 3 (two-phase kinetic emission model), Chapter 6 (chemometric and machine-learning classification), Chapter 7 (Yarn Shedding Index and Safe and Sustainable by Design), Chapter 8 (surface-mediated enzymatic depolymerization), Chapter 11 (consumption-velocity dynamics).

Master Bibliography

The Textile Matrix: Modeling, Measuring, and Mitigating Textile Microfiber Pollution

Alimba, C. G., & Faggio, C. (2019). Microplastics in the marine environment: A review of the sources, occurrence, fate, and effects. Environmental Toxicology and Pharmacology, 72, 103251.

Amato-Lourenço, L. F., et al. (2021). Presence of airborne microplastics in human lung tissue. Journal of Hazardous Materials, 416, 126124.

Andrady, A. L. (2017). Microplastics in the marine environment. Marine Pollution Bulletin, 115(1-2), 5–22.

Bautista-Zamudio, P.A., et al. (2023). Biodegradation of plastics by white-rot fungi: A review. Science of the Total Environment.

Boccia, P., et al. (2024). Potential Effects of Environmental and Occupational Exposure to Microplastics. Toxics, 12(5), 320.

Boucher, J., & Friot, D. (2017). Primary microplastics in the oceans: a global evaluation of sources. IUCN.

Campanale, C., et al. (2020). A detailed review study on potential effects of microplastics and additives of concern on human health. International Journal of Environmental Research and Public Health, 17(4), 1212.

Chan, C. M., et al. (2019). Polyhydroxyalkanoates (PHAs) as sustainable alternatives to synthetics. European Polymer Journal, 116, 406–419.

Choi, S., et al. (2021). Analysis of Microplastics Released from Plain Woven Fabrics with Different Yarn Constructions. Polymers.

Cole, M., et al. (2011). Microplastics as contaminants in the marine environment: A review. Marine Pollution Bulletin, 62(12), 2588–2597.

Cui, H., et al. (2022). Study on the Relationship between Textile Microplastics and Fabric Structure/Finishing. Polymers, 14(23), 5309.

De Falco, F., et al. (2020). Microfiber Release to Water, Via Laundering, and to Air, Via Everyday Use: A Comparison between Polyester Clothing with Differing Textile Parameters. Environmental Science & Technology.

Dehaut, A., et al. (2016). Microplastics in seafood: Benchmark protocol for their extraction and characterization. Environmental Pollution, 215, 223–233.

Dris, R., et al. (2017). A first overview of textile fibers, including microplastics, in indoor and outdoor environments. Environmental Pollution, 221, 453–458.

Ellen MacArthur Foundation. (2015). Circularity Indicators: An Approach to Measuring Circularity. Cowes, UK.

European Commission. (2022/2024). Ecodesign for Sustainable Products Regulation (ESPR) and Digital Product Passport framework. Brussels.

European Commission. (2018). A European Strategy for Plastics in a Circular Economy. COM(2018) 28 final. Brussels.

European Commission. (2020/2022). Safe and Sustainable by Design (SSbD) Framework for Chemicals and Materials. JRC Technical Reports.

Fischer, M., & Scholz-Böttcher, B.M. (2017). Simultaneous trace identification and quantification of common types of microplastics in environmental samples by pyrolysis-gas chromatography–mass spectrometry. Environmental Science & Technology, 51(9), 5052–5060.

Fitch, S., et al. (2025). Systematic review of potential developmental and reproductive toxicity of microplastics. Toxicological Sciences, 207(2), 289–305.

Fletcher, K. (2016). Craft of Use: Post-Growth Fashion. Routledge.

Gayathri, P.V., et al. (2023). Advanced oxidation processes for the degradation of microplastics: A review. Water Environment Journal.

Geyer, R., Jambeck, J. R., & Law, K. L. (2017). Production, use, and fate of all plastics ever made. Science Advances, 3(7), e1700782.

Hazlehurst, A., et al. (2024). Investigating the influence of yarn characteristics on microfibre release during laundering. Frontiers in Environmental Science.

Henry, B., Laitala, K., & Klepp, I. G. (2019). Microfibres from apparel and home textiles. Sustainable Production and Consumption, 20, 263–275.

Hidalgo-Ruz, V., et al. (2012). Microplastics in the marine environment: A review of the methods used for identification and quantification. Environmental Science & Technology, 46(6), 3060–3075.

Hufnagl, B., de Frutos, D., & Primpke, S. (2022). GEPARD: An open-source software tool for automated database coordinate mapping in microplastics Raman/FTIR spectroscopy. Journal of Open Source Software, 7(71), 4102.

Kang, H., et al. (2021). Impact of Chitosan Pretreatment to Reduce Microfibers Released from Synthetic Textiles during Laundering. Water, 13(18), 2480.

Khanam, M.M., et al. (2025). Advances in machine learning for the detection and classification of microplastics. Frontiers in Environmental Science.

Kiekens, P., & El-Ghalia, M. (2021). Microfiber shedding behaviors on chitosan-finished PET fabrics. Textile Research Journal, 91(13-14), 1544–1556.

Kumar, J., et al. (2025). Trapping tiny pollutants: SERS-driven strategies for microplastic detection and characterization. Journal of Hazardous Materials.

Langella, C., et al. (2022). Electrofluidodynamic coating of biodegradable polymers (PLA/PBSA) on nylon fabrics. Polymer Degradation and Stability, 196, 109–123.

Lant, N. J., et al. (2020). Microfiber release from real soiled consumer laundry and the impact of fabric care products. PLOS ONE, 15(6), e0233332.

Leslie, H. A., et al. (2022). Discovery and quantification of plastic particle pollution in human blood. Environment International, 163, 107199.

Li, C., et al. (2025). Potential planetary health impacts of the airborne plastisphere. One Earth. https://doi.org/10.1016/j.oneear.2025.101446

Löder, M.G.J., et al. (2017). Enzymatic purification of microplastics in environmental samples. Environmental Science & Technology, 51(24), 14283–14292.

Mintenig, S.M., et al. (2017). Identification of microplastic in effluents of waste water treatment plants using focal plane array-based micro-Fourier-transform infrared imaging. Water Research, 108, 365–372.

Muthusamy, K., Ibrahim, S., & Sarlin, E. (2023). Quantifying temperature boundaries and microfiber release kinetics during commercial denim washing. Environmental Pollution, 318, 120890.

Napper, I. E., & Thompson, R. C. (2016). Release of synthetic microplastic fibres from domestic washing machines. Marine Pollution Bulletin, 112(1-2), 39–45.

Niinimäki, K., et al. (2020). The environmental price of fast fashion. Nature Reviews Earth & Environment, 1(4), 189–200.

OECD. (2022). Global Plastics Outlook: Policy Scenarios to 2060. OECD Publishing, Paris. https://doi.org/10.1787/aa1edf33-en

OECD. (2023). New Aspects of EPR: Extending producer responsibility to additional product groups and environmental impacts. OECD Environment Working Papers No. 217. https://doi.org/10.1787/8f8e3c8f-en

Otsu, N. (1979). A threshold selection method from gray-level histograms. IEEE Transactions on Systems, Man, and Cybernetics, 9(1), 62–66.

Papachristou, E., & Bilalis, N. (2016). Can 3D virtual prototyping replace physical prototyping in apparel industry? International Journal of Fashion Design, Technology and Education, 9(1), 25–34.

PlasticsEurope. (2025). Plastics – the Fast Facts 2025. PlasticsEurope Market Research Group.

Prata, J. C., et al. (2020). Environmental exposure to microplastics: An overview on possible human health effects. Science of the Total Environment, 702, 134455.

Primpke, S., et al. (2017). An automated approach for microplastics analysis using focal plane array (FPA) FTIR microscopy and image analysis. Analytical Methods, 9(9), 1499–1511.

Ragusa, A., et al. (2021). Plasticenta: First evidence of microplastics in human placenta. Environment International, 146, 106274.

Rahman, A., et al. (2021). Potential human health risks due to environmental exposure to microplastics and knowledge gaps: A scoping review. Science of the Total Environment, 757, 143872.

Rochman, C. M., et al. (2013). Policy: Classify plastic waste as hazardous. Nature, 494(7436), 169–171.

Schwabl, P., et al. (2019). Detection of various microplastics in human stool: A prospective case series. Annals of Internal Medicine, 171(7), 453–457.

Shen, L., Worrell, E., & Patel, M. K. (2010). Environmental impact assessment of man-made cellulose fibres. Resources, Conservation and Recycling, 55(2), 125–133.

Tao, D., et al. (2022). Microfibers Released into the Air from a Household Tumble Dryer. Environmental Science & Technology Letters.

Textile Exchange. (2025). Materials Market Report 2025. Textile Exchange.

The Microfibre Consortium. (n.d.). Glossary of terms relating to microfibres and fibre fragments. https://www.microfibreconsortium.com/glossary

ThredUp. (2024). Resale Report 2024. https://www.thredup.com/resale-report (Accessed June 2026).

Tomasino, C. (1992). Chemistry & technology of fabric preparation & finishing. North Carolina State University.

Tournier, V., et al. (2020). An engineered PET depolymerase to break down and recycle plastic bottles. Nature, 580, 216–219.

United Nations Environment Assembly. (2022–ongoing). Intergovernmental Negotiating Committee on Plastic Pollution (Global Plastics Treaty negotiations). Nairobi.

Vassilenko, E., et al. (2021). Domestic laundry and microfiber pollution: Exploring fiber fragmentation and novel mitigation strategies. Environmental Science & Technology.

Vethaak, A. D., & Leslie, H. A. (2016). Plastic debris is a human health issue. Environmental Science & Technology, 50(13), 6825–6826.

Vial, J., Dupont, L., & Mercier, F. (2023). Quantification of microplastics in bottled drinking water using Nile Red staining, automated atlas mapping, and random forest models. Food Chemistry, 402, 134–145.

Wei, R., & Zimmermann, W. (2017). Microbial enzymes for the recycling of recalcitrant petroleum-based plastics: How far are we? Microbial Biotechnology, 10(6), 1308–1322.

World Health Organization (WHO). (2019). Dietary and inhalation exposure to microplastics in the human environment: Risk characterization. Geneva.

Wright, S. L., & Kelly, F. J. (2017). Plastic and human health: A micro issue? Environmental Science & Technology, 51(12), 6634–6647.

Yoshida, S., et al. (2016). A bacterium that degrades and assimilates poly(ethylene terephthalate). Science, 351(6278), 1196–1199.

Zettler, E.R., Mincer, T.J., & Amaral-Zettler, L.A. (2013). Life in the "Plastisphere": Microbial Communities on Plastic Marine Debris. Environmental Science & Technology, 47(13), 7137–7146.

Murden, S., et al. (2024). Low-cost, high-throughput quantification of microplastics released from textile wash tests: introducing the Fibre Fragmentation Scale. Cambridge Prisms: Plastics.

Glossary of Key Terms

This glossary defines key technical, scientific, and policy terms used throughout The Textile Matrix. Terms are organized alphabetically for quick reference.

ALDFG (Abandoned, Lost, or otherwise Discarded Fishing Gear): Fishing nets, lines, and traps that are no longer under the control of fishers. A major source of marine plastic pollution and microplastic generation through fragmentation.

Biofouling: The accumulation of microorganisms, plants, algae, and small animals on submerged surfaces, including microplastics. Biofouling increases the density of floating plastics, causing them to sink and altering their environmental fate and toxicity.

Circular Economy: An economic model that aims to eliminate waste and keep materials in use for as long as possible through reuse, repair, remanufacturing, and recycling. Contrasts with the linear 'take-make-dispose' model.

Digital Product Passport (DPP): A digital record containing information about a product's materials, components, environmental impact, and end-of-life instructions. Mandated under the EU Ecodesign for Sustainable Products Regulation (ESPR) for textiles.

Eco-corona: A layer of biomolecules (proteins, polysaccharides, lipids) that rapidly adsorbs onto the surface of micro- and nanoplastics when they enter aqueous environments. The eco-corona alters particle identity, aggregation behavior, and biological interactions.

Extended Producer Responsibility (EPR): A policy approach in which producers are given significant responsibility—financial and/or physical—for the treatment or disposal of post-consumer products. Modulated EPR fees can incentivize better design.

Fragmentation: The physical breakdown of larger plastic items into smaller particles (microplastics and nanoplastics) through mechanical stress, UV photo-oxidation, hydrolysis, and biological activity. The dominant pathway for secondary microplastic formation.

GEPARD (Gepard Automated Raman/FTIR Diagnostics): An open-source software tool and analytical workflow that uses coordinate-targeted spectroscopy to dramatically reduce the time required for microplastic identification on filter membranes.

In silico fabric screening: The use of computational models to predict the shedding behavior of a textile before physical manufacture, enabling rapid ranking of design options for low-shedding performance.

Microfiber: A slender, elongated microplastic particle, typically with a high aspect ratio (>3:1), originating primarily from the shedding of synthetic textiles during laundering, wear, and weathering.

Microplastic: Water-insoluble polymeric particles with a longest dimension between 1 μm and 5 mm. Includes fragments, fibers, films, foams, and pellets. Defined by ISO and widely adopted regulatory frameworks.

Multi-scale modeling: An integrative computational approach that links models operating at different scales (material, event, product-lifetime, and population) to predict overall environmental emissions from design decisions through to aggregate flux.

Nanoplastic: Polymeric particles with at least one dimension smaller than 1 μm (1000 nm). Capable of crossing biological membranes, including the blood-brain and placental barriers. More reactive and bioavailable than microplastics.

Physics-informed machine learning: A hybrid modeling approach that combines mechanistic physical models with data-driven machine learning. The mechanistic component provides known physical structure, while the machine-learning component learns residuals or unmodeled effects.

Plastisphere: The distinct microbial community that colonizes plastic debris in aquatic environments. The plastisphere can differ significantly from surrounding microbial communities and may facilitate pathogen transport and plastic degradation.

Primary Microplastic: Microplastics that are manufactured at microscopic size for specific purposes (e.g., microbeads in cosmetics, industrial abrasives, or pre-production pellets/nurdles). Enter the environment directly.

Safe and Sustainable by Design (SSbD): A European Commission framework that integrates safety, sustainability, and circularity considerations into the earliest stages of chemical and material development. Aims to minimize hazard and exposure throughout the lifecycle.

Secondary Microplastic: Microplastics formed through the environmental fragmentation of larger plastic items. Constitute the majority (70–90%) of environmental microplastic burden. Textile-derived microfibers are a major source of secondary microplastics.

Textile-lifecycle digital twin: A dynamic computational model of a garment or product line that integrates multi-scale emission predictions and updates forecasts as design, consumption, and end-of-life parameters change over time.

Transfer function (predictive): A mathematical or statistical relationship that maps measurable textile descriptors (polymer type, yarn twist, fabric construction, etc.) to predicted fragmentation or emission behavior across the product lifecycle.

Two-phase kinetic emission model: A mathematical model describing microfiber release during laundering as the sum of a rapidly decaying transient term (pre-existing loose fragments) and a continuous generation term (new fragments produced during washing).

Yarn Shedding Index: A dimensionless index that estimates a fabric's intrinsic shedding propensity from structural parameters such as fiber type, staple length, yarn twist, and fabric construction. Operates at the material design scale.

Subject Index

This index provides references to key concepts, materials, processes, technologies, and policy instruments discussed in The Textile Matrix. Page numbers refer to the print edition.

www.ingramcontent.com/pod-product-compliance
Lightning Source LLC
La Vergne TN
LVHW051055210926
848486LV00002B/115
9781764624954